PLEASE STAMP DATE DUE, BOTH BELOW AND ON CARD

DATE DUE	DATE DUE	DATE DUE	DATE DUE
FEB - 1 2012			

GL-15

Advances in Geophysical and Environmental Mechanics and Mathematics

Series Editor: Professor Kolumban Hutter

For further volumes:
http://www.springer.com/series/7540

Board of Editors

Aeolean Transport, Sediment Transport, Granular Flow
Prof. Hans Herrmann
Institut für Baustoffe
Departement Bau, Umwelt und Geomatik
HIF E 12/ETH Hönggerberg
8093 Zürich, Switzerland
hjherrmann@ethz.ch

Avalanches, Landslides, Debris Flows, Pyroclastic Flows, Volcanology
Prof E. Bruce Pitman
Department of Mathematics
University of Buffalo
Buffalo, N. Y. 14260, USA
Pitman@buffalo.edu

Hydrological Sciences
Prof. Vijay P. Singh
Water Resources Program
Department of Civil and Environmental Engineering
Louisiana State University
Baton Rouge, LA 70803-6405, USA

Nonlinear Geophysics
Prof. Efim Pelinovsky
Institute of Applied Physics
46 Uljanov Street
603950 Nizhni Novgorod, Russia
enpeli@mail.ru

Planetology, Outer Space Mechanics
Prof Heikki Salo
Division of Astronomy
Department of Physical Sciences
University of Oulu
90570 Oulu, Finnland

Glaciology, Ice Sheet and Ice Shelf Dynamics, Planetary Ices
Prof. Dr. Ralf Greve
Institute of Low Temperature Science
Hokkaido University
Kita-19, Nishi-8, Kita-ku
Sapporo 060-0819, Japan
greve@lowtem.hokudai.ac.jp
http://wwwice.lowtem.hokudai.ac.jp/~greve/

Thomas Stocker

Introduction to Climate Modelling

Thomas Stocker
Universität Bern
Abt. Klima- und Umweltphysik
Sidlerstrasse 5
3012 Bern
Switzerland
stocker@climate.unibe.ch

Advances in Geophysical and Environmental Mechanics and Mathematics
ISSN 1866-8348   e-ISSN 1866-8356
ISBN 978-3-642-00772-9   e-ISBN 978-3-642-00773-6
DOI 10.1007/978-3-642-00773-6
Springer Heidelberg Dordrecht London New York

Library of Congress Control Number: 2011929501

© Springer-Verlag Berlin Heidelberg 2011

This work is subject to copyright. All rights are reserved, whether the whole or part of the material is concerned, specifically the rights of translation, reprinting, reuse of illustrations, recitation, broadcasting, reproduction on microfilm or in any other way, and storage in data banks. Duplication of this publication or parts thereof is permitted only under the provisions of the German Copyright Law of September 9, 1965, in its current version, and permission for use must always be obtained from Springer. Violations are liable to prosecution under the German Copyright Law.

The use of general descriptive names, registered names, trademarks, etc. in this publication does not imply, even in the absence of a specific statement, that such names are exempt from the relevant protective laws and regulations and therefore free for general use.

Cover design: deblik, Berlin

Printed on acid-free paper

Springer is part of Springer Science+Business Media (www.springer.com)

*Dedicated in memoriam to
Daniel Gordon Wright (1952–2010)*

Preface

These Lecture Notes have been developed during the past few years and form the basis of a one-semester course at the Master level offered by the Physics Institute and the Oeschger Centre for Climate Change Research at the University of Bern in their Master Courses in Physics and in Climate Sciences, respectively. The goal of this course is to provide the students with a first and basic approach to the subject of climate modelling.

Climate modelling has become an important pillar in physically based climate research. Climate research is active in both directions on the time axis, i.e., both reconstruction of *past climate change* as well as an estimate of *future climate change*. As past climate change is reconstructed from a growing number of paleoclimatic archives with increasing detail and regional coverage, a mechanistic understanding of these changes requires climate models. Only through climate models are we able to test hypotheses and to make predictions of climate changes in time windows yet to be accessed by new paleoclimatic records. The opposite direction of the time axis, the projection of the future, can only be accessed by performing experiments using climate models which enable the numerical computation of sets of equations and parameterisations that describe the various inter-connected processes in the Earth system and their combination with scenarios of future human development, various paths of emissions of greenhouse gases, and natural physical changes in the climate system.

For some years now, I have noted a tendency in the climate modelling community that complex models are being increasingly used for experiments by many students and researchers who have never been directly involved in their development. In the worst case, climate models are used as black boxes much like "clickable software" so familiar to everybody. What is more and more missing is the direct contact of the students with all the necessary steps of the subject matter: (a) formulation of the climate problem in terms of equations which describe the fundamental physical processes under investigation; (b), formulation of parameterisations in order to account for unresolved processes; (c), selection of numerical schemes to solve the equations; (d), critically testing, tuning and further developing the climate model; and (e), applying the climate model to problems suitable for the chosen structure of the model. Instead, students are too often confronted with a ready-to-use climate model system which is then uncritically applied to the problem.

These Lecture Notes should make a contribution to provide the students with some basic knowledge of climate model development. The approach that I have chosen is a close combination of three elements: (a) fundamental climate processes; (b), their formulation in simplified physical models including parameterisations; and (c), their numerical solution. These three elements are a recurring theme in climate model development across the entire model hierarchy ranging from models of reduced complexity to comprehensive Earth System models. Therefore, these notes are hopefully serving as a first, but still quantitative, entry point for students interested in the modelling aspects of climate science. It is aimed at students of Earth Sciences and Physics, as well as mathematically interested students of Climate Geology and Climate Geography.

There are clearly limitations inherent to a one-semester course. Only the most fundamental climate processes are presented and some basic numerical concepts are explained here. The students are then directed to more advanced texts and material which should be better accessible after digestion of these Notes.

Many colleagues and individuals have contributed knowingly and unknowingly in one way or another to these Lecture Notes over the past years. I would like to thank my PhD supervisor Kolumban Hutter at ETH Zürich, who taught me how to analyse physical systems with rigor in the mid 1980s. His ability to capture complex processes in mathematical descriptions is unsurpassed. He also encouraged me to submit my notes to this book series. Lawrence Mysak had a major influence on me and was a generous and stimulating host during my postdoctoral years at McGill University where, together with the late Dan Wright, we laid the foundation of climate modelling of intermediate complexity. Lawrence's enthusiasm for science is simply contagious and his support of young scientists is remarkable. Without Wally Broecker of Columbia's Lamont Doherty Earth Observatory, the application and continuous development and expansion of climate models of intermediate complexity would not have been possible. He constantly pushed the boundaries of what we should do with these models. Keeping a natural skepticism towards models, Wally finally accepted what we learned from models: There is no physical basis that abrupt climate change would be associated with worldwide cooling or warming, a debate that we passionately carried out in the mid 1990s when the paleo-data were sending apparently conflicting messages.

In the past 18 years, the colleagues at the Division of Climate and Environmental Physics of the University of Bern have been a constant source of inspiration, all PhD students and postdocs, and in particular Andreas Schmittner, Niki Gruber, Joël Hirschi, Kasper Plattner, Reto Knutti and Stefan Ritz. The postdoctoral fellows Olivier Marchal, Christoph Appenzeller, Gilles Delaygue, Masa Yoshimori, Neil Edwards, and Mark Siddall have also made major contributions to the development of physical–biogeochemical climate models of reduced complexity, as well as to their application. The latter would not have been possible without the insights of my colleague Fortunat Joos.

I gratefully acknowledge the contributions of Christoph Raible at various stages of the production of these Lecture Notes and during the joint teaching of this course. Beni Stocker provided a first version of the English translation, and Flavio

Lehner produced custom-made figures from the ERA-40 and NCEP data sets. Beat Ihly has masterfully translated the original manuscript into LaTeX, checked critically all derivations and redrafted many figures. He also significantly expanded Sects. 3.4, 6.4 and parts of Sect. 6.8 to make them more accessible for the students. Remaining errors, naturally, are my own and sole responsibility.

I wish to dedicate these Lecture Notes to the memory of the late Dr. Daniel G. Wright (1952–2010), Senior Research Scientist of Bedford Institute of Oceanography, Dartmouth, Canada. Dan was a dear friend and my closest collaborator during my formative years in the early 1990s. I was privileged to develop with Dan a vision of climate models of intermediate complexity and, with him, to take the first steps towards achieving this goal. The development of the zonally averaged, coupled climate model, the "Wright–Stocker model", or the later "Bern 2.5-dimensional model", was a very exciting and rewarding experience which I owe to Dan. His sharp insight into the physics of the ocean and his dedication to the development of simplified, yet physically-based, models have changed my thinking.

Finally, without the constant support and love by my family, my wife Maria-Letizia, and the daughters Francesca and Anna-Maria, life would be truly one-dimensional.

Bern, January 2011 *Thomas Stocker*

Contents

1 **Introduction** 1
   1.1 Goals of these Notes 1
   1.2 The Climate System 3
       1.2.1 Components of the Climate System 3
       1.2.2 Global Radiation Balance of the Climate System 5
   1.3 Purpose and Limitations of Climate Modelling 5
   1.4 Historical Development... 10
   1.5 Some Current Examples in Climate Modelling 15
       1.5.1 Simulation of the Twentieth Century with the Goal to Quantify the Link Between Increases in Atmospheric CO_2 Concentrations and Changes in Temperature 15
       1.5.2 Decrease in Arctic Sea Ice Cover Since Around 1960 16
       1.5.3 Summer Temperatures in Europe Towards the End of the Twenty-First Century 17
       1.5.4 CO_2 Emissions Permitted for Prescribed Atmospheric Concentration Paths 18
       1.5.5 Prediction of the Weak El Niño of 2002/2003 19
   1.6 Conclusions 21

2 **Model Hierarchy and Simplified Climate Models** 25
   2.1 Hierarchy of Physical Climate Models................................... 25
   2.2 Point Model of the Radiation Balance 34
   2.3 Numerical Solution of an Ordinary Differential Equation of First Order 37
   2.4 Climate Sensitivity and Feedbacks...................................... 41
       2.4.1 Ice-Albedo Feedback 43
       2.4.2 Water Vapour Feedback 45
       2.4.3 Cloud Feedback... 45
       2.4.4 Lapse Rate Feedback 47
       2.4.5 Summary and Conclusion Regarding Feedbacks 48

Contents

3 Describing Transports of Energy and Matter 53
  3.1 Diffusion 53
  3.2 Advection 55
  3.3 Advection-Diffusion Equation and Continuity Equation 56
  3.4 Describing Small- and Large-Scale Motions 57
  3.5 Solution of the Advection Equation 61
    3.5.1 Analytical Solution 61
    3.5.2 Numerical Solution 63
    3.5.3 Numerical Stability, CFL Criterion 64
  3.6 Further Methods for the Solution of the Advection Equation 68
    3.6.1 Euler Forward in Time, Centered in Space (FTCS) 68
    3.6.2 Euler Forward in Time, Upstream in Space (FTUS) 69
    3.6.3 Implicit Scheme 70
    3.6.4 Lax Scheme 72
    3.6.5 Lax–Wendroff Scheme 74
  3.7 Numerical Solution of the Advection-Diffusion Equation 75
  3.8 Numerical Diffusion 76

4 Energy Transport in the Climate System and Its Parameterisation 79
  4.1 Basics 79
  4.2 Heat Transport in the Atmosphere 79
  4.3 Meridional Energy Balance Model 83
  4.4 Heat Transport in the Ocean 85

5 Initial Value and Boundary Value Problems 91
  5.1 Basics 91
  5.2 Direct Numerical Solution of Poisson's Equation 92
  5.3 Iterative Methods 94
    5.3.1 Methods of Relaxation 94
    5.3.2 Method of Successive Overrelaxation (SOR) 95

6 Large-Scale Circulation in the Ocean 97
  6.1 Material Derivative 97
  6.2 Equation of Motion 98
  6.3 Continuity Equation 101
  6.4 Special Case: Shallow Water Equations 101
  6.5 Different Types of Grids in Climate Models 105
  6.6 Spectral Models 108
  6.7 Wind-Driven Flow in the Ocean (Stommel Model) 110
    6.7.1 Determination of the Stream Function 112
    6.7.2 Determination of the Water Surface Elevation 114
  6.8 Potential Vorticity: An Important Conserved Quantity 117

Contents

7 Large-Scale Circulation in the Atmosphere ... 123
   7.1 Zonal and Meridional Circulation ... 123
   7.2 The Lorenz–Saltzman Model ... 128

8 Atmosphere–Ocean Interactions ... 137
   8.1 Coupling of Physical Model Components ... 137
   8.2 Thermal Boundary Conditions ... 137
   8.3 Hydrological Boundary Conditions ... 142
   8.4 Momentum Fluxes ... 144
   8.5 Mixed Boundary Conditions ... 145
   8.6 Coupled Models ... 146

9 Multiple Equilibria in the Climate System ... 151
   9.1 Abrupt Climate Change Recorded in Polar Ice Cores ... 151
   9.2 The Bipolar Seesaw ... 152
   9.3 Multiple Equilibria in a Simple Atmosphere Model ... 156
   9.4 Multiple Equilibria in a Simple Ocean Model ... 157
   9.5 Multiple Equilibria in Coupled Models ... 159
   9.6 Concluding Remarks ... 163

References ... 165

Index ... 171

Acronyms

AD	Anno Domini
AGCM	Atmospheric General Circulation Model
AMIP	Atmospheric Modelling Intercomparison Project
AOGCM	Atmosphere/Ocean General Circulation Model
C^4MIP	Coupled Carbon Cycle Climate Modelling Intercomparison Project
CFL	Courant–Friedrichs–Lewy
CMIP	Coupled Climate Modelling Intercomparison Project
COADS	Comprehensive Ocean Atmosphere Data Set
CPC	Climate Prediction Center (NOAA, NWS)
CTCS	Centered differences in Time, Centered differences in Space
EBM	Energy Balance Model
EMIC	Earth System Model of Intermediate Complexity
ENIAC	Electronic Numerical Integrator and Computer
ENSO	El Niño-Southern Oscillation
EPICA	European Project for Ice Coring in Antarctica
ERBE	Earth Radiation Balance Experiment
ESRL	Earth System Research Laboratory (NOAA)
FCT	Flux-Corrected Transport
FTCS	Forward differences in Time, Centered differences in Space
FTUS	Forward differences in Time, Upstream differences in Space
GCM	General Circulation Model
GRIP	Greenland Ice Core Project
IGES	Institute of Global Environment and Society
IPCC	Intergovernmental Panel on Climate Change
ITCZ	Intertropical Convergence Zone
KUP	Division of Climate and Environmental Physics (Abteilung für Klima- and Umweltphysik) at the University of Bern
MIS	Marine Isotope Stage
MOC	Meridional Overturning Circulation
MPI	Max-Planck-Institut
NASA	National Aeronautics and Space Administration
NASA-GISS	NASA Goddard Institute for Space Studies
NCAR	National Center for Atmospheric Research

NCEP	National Centers for Environmental Prediction
NOAA	National Oceanic and Atmospheric Administration
NWS	National Weather Service (NOAA)
OCMIP	Ocean-Carbon Cycle Modelling Intercomparison Project
ODE	Ordinary Differential Equation
OGCM	Ocean General Circulation Model
OMIP	Ocean Modelling Intercomparison Project
PCMDI	Program for Climate Model Diagnosis and Intercomparison
PDE	Partial Differential Equation
PMIP	Paleoclimate Modelling Intercomparison Project
QG	Quasi-Geostrophic
SE	Stationary Eddy
SOR	Successive Overrelaxation
SRES	Special Report on Emissions Scenarios (IPCC)
SSS	Sea Surface Salinity
SST	Sea Surface Temperature
TE	Transient Eddy
THC	Thermohaline Circulation

List of Frequently Occurring Symbols

$\vec{\nabla}$	Gradient operator	$\vec{\nabla} = \left(\frac{\partial}{\partial x}, \frac{\partial}{\partial y}, \frac{\partial}{\partial z}\right)$
$\vec{\nabla}^2$	Laplace operator	$\vec{\nabla}^2 = \frac{\partial^2}{\partial x^2} + \frac{\partial^2}{\partial y^2} + \frac{\partial^2}{\partial z^2}$
$\frac{D}{Dt}$	Material derivative	$\frac{D}{Dt} = \frac{\partial}{\partial t} + u\frac{\partial}{\partial x} + v\frac{\partial}{\partial y} + w\frac{\partial}{\partial z}$
α	Albedo or volume coefficient of expansion	
β	$= df/dy$	
γ	Lapse rate	
ε	Emissivity	
ζ	Relative vorticity	
η	Water surface elevation	
θ	Temperature (Lorenz–Saltzman model)	
κ	Thermal diffusivity	
λ	Geographical longitude; climate feedback parameter; wave length	
ν	Kinematic viscosity; frequency	
ξ	Residual (SOR)	
ρ	Mass density	
σ	Stefan–Boltzmann constant	
τ	Time scale; time interval; friction force	
φ	Geographical latitude	
Ψ	Stream function	
Ω	Domain of a mathematical function	
$\vec{\Omega}$	Angular velocity of Earth's rotation	
Ω	Magnitude of angular velocity of Earth's rotation	
ω	Angular velocity; angular frequency; overrelaxation parameter (SOR)	
$\vec{a}$	Acceleration	
$\vec{a}_I$	Inertial acceleration	
a	Magnitude of acceleration	
a_x, a_y, a_z	Components x, y and z of acceleration	
A	Eddy viscosity	

C	Concentration; density
c; c_V, c_p	Specific heat capacity; specific heat capacities at constant volume and at constant pressure
c_D	Transfer coefficient for momentum (dimensionless)
D	Diffusion constant, diffusion coefficient, diffusivity; transfer coefficient for the sensible heat flux
f	Coriolis parameter
$\vec{F}$	Flux or flux density
F	Magnitude of flux or flux density
$F^{O \to A}$	Magnitude of the flux from the ocean to the atmosphere
$\vec{g}$	Gravity acceleration or free-fall acceleration
g	Magnitude of gravity acceleration or free-fall acceleration
H, h	Layer thickness
k	Wave number
K	Eddy diffusion constant (eddy diffusivity)
L	Angular momentum; specific latent heat
$\vec{M}$	Mass transport
M_x, M_y	Horizontal components of the mass transport
n	Particle density
p	Pressure
q	Specific humidity; surface flow
Q^{short}	Solar radiation (mainly short-wave)
R	Earth's radius
R^*	Specific gas constant of air
s	Climate sensitivity
S	Radiation; salinity
S_0	Solar constant
t	Time
Δt	Time step, time interval
T	Temperature; period
T^*	Restoring temperature
$\vec{u}$	Velocity
u	Magnitude of velocity
u, v, w	Components x, y and z of velocity
V	Volume
x, y, z	Coordinates x, y and z
$\Delta x, \Delta y, \Delta z$	Distances in space; grid spacing

Chapter 1
Introduction

1.1 Goals of these Notes

These Lecture Notes form the basis of a one-semester course taught at the Physics Institute and the Oeschger Centre of Climate Change Research of the University of Bern. The *main goals* of this course are:

1. To introduce the students to the physical basis and the mathematical description of the different *components of the climate system*.
2. To provide the students with a first approach to the *numerical solution* of ordinary and partial differential equations using examples from climate modelling.

The combination of these two goals in one course should enable the students, who are increasingly becoming users of climate models and are missing the direct contact and developmental involvement with climate models, to gain an insight into the construction of climate model components, the nature of parameterisations and some of the potential pitfalls of numerical computation in the context of climate modelling. The present Lecture Notes aim to achieve this by presenting and illustrating a few simple and basic examples of how different components of the Earth system are simulated, including the processes governing their dynamics and their relevance for past and future climate change.

Numerical climate models enable a physically based estimate of the range of future climate change. These models, which rest on the fundamental laws of physics and chemistry (conservation of energy, mass, momentum, etc.), are invaluable in providing scientific information towards political and societal decision making. When the effects of a doubling of the atmospheric CO_2 concentration, as it is expected around the year 2050, and other changes in the atmospheric composition have to be evaluated, only numerical climate models can generate a well-founded quantitative answer.

Climate models bring together findings of many disciplines in natural sciences. The understanding of dynamical processes in the atmosphere and the ocean is crucial for its modelling. Fluid dynamics in a rotating frame of reference (geophysical fluid dynamics) plays a major role. The resulting partial differential equations need to be solved with calculation schemes: a problem for numerical mathematics. As in each model representation of natural systems, there are processes that cannot be simulated because they are insufficiently understood or because they occur on temporal or spatial scales which the model cannot capture. Therefore, parameterisations are formulated, some of which will be presented in this book.

There are some helpful textbooks on the topic of climate and climate modelling:

- Peixoto J.P., Oort, A.H., 1992, *Physics of Climate*, 2nd ed., American Institute of Physics, 520 pp.

  Very clear and detailed introduction into the physical basis of the climate system and its different components (Atmosphere, Ocean, Ice). Good presentation of the climatology of important quantities. The aspect of climate models, however, is treated only briefly.

- *Climate System Modeling*, 1992, K.E. Trenberth (Editor), Cambridge University Press, 788 pp.

  Coherent collection of overview articles on climate modelling, particularly the different components, biogeochemical cycles included, presented in four parts: basic processes, modelling and parameterisation, coupling of the different systems and applications. In some cases no longer up-to-date.

- McGuffie K., A. Henderson-Sellers, 2005, *A Climate Modelling Primer*, 3rd ed., John Wiley, 296 pp.

  Introduction into the hierarchy of models and formulations including examples and programs.

- Washington W.M., C.L. Parkinson, 2005, *An Introduction to Three-Dimensional Climate Modeling*, University Science Books, 354 pp.

  Clear presentation of the physics of the different system components, not as detailed as Peixoto & Oort (1992), but closer related to modelling. Many parameterisations are described. An updated classic of 1986.

- *Ocean Circulation and Climate: Observing and Modelling the Global Ocean*, 2001, G. Siedler, J. Church, J. Gould (Eds.), International Geophysics Series 77, Academic Press, 2001, 715 pp.

  Very good overview of research in oceanography on a global scale. Excellent figures.

- Houghton J., 2002, *The Physics of Atmospheres*, 3rd ed., Cambridge University Press, 320 pp.

  Basic and comprehensive presentation of the physics of the atmosphere (radiation, clouds, circulation), with an overview of climate change, climate models and predictability.

- Hartmann D.L., 1997, *Global Physical Climatology*, Academic Press, 411 pp.

  Very clear and rigorous introduction to the physics of ocean and atmosphere and a physically-based discussion on climate variability and climate change.

- *Climate Change 2007: The Physical Science Basis. Contribution of Working Group I to the Fourth Assessment Report of the Intergovernmental Panel on Climate Change.* S. Solomon et al. (Eds.), Cambridge University Press.

  Comprehensive presentation of the latest research of climate sciences as of the year 2006, referring to the question of climate change. The complete report is available under http://www.ipcc.ch.

- Houghton J., 2009, *Global Warming: The Complete Briefing*, 4th ed., Cambridge, 456 pp.
  Excellent and uptodate overview of the science knowledge regarding global warming and consequences. Sir John Houghton was Co-Chair of IPCC for the Second and Third Assessment Reports of the Intergovernmental Panel on Climate Change published in 1995 and 2001, respectively.

Some books on the basics of numerical solutions of problems in mathematical physics:

- Schwarz, H.R., 2004, *Numerische Mathematik 7. Auflage*, Teubner, 653 pp.
  German. Good introduction into the different numerical methods, interpolation, integration and solution of partial differential equations. Numerous examples.

- Press W.H., S.A. Teukolsky, W.T. Vetterling, B.P. Flannery, 1992, *Numerical Recipes in Fortran (Volumes 1 & 2)*, Cambridge, 963 pp. (Volume 2 for Fortran 90, 1996).
  Large collection of numerical schemes in different programming languages. Schemes are explained briefly and succinctly. Their good and bad properties are discussed. Must be part of the library of every modeler. The newest edition (third edition, 2007) is written in C++.

- Krishnamurti T.N., L. Bounoua 1996, *An Introduction to Numerical Weather Prediction Techniques*. CRC Press, 304 pp.
  Comprehensive explanation of different solving schemes and parameterisations which are used in atmospheric circulation models.

- Haltiner, G.J., R.T. Williams, 1980, *Numerical Prediction and Dynamic Meteorology*. Wiley, 477 pp.
  Advanced text with many derivations of numerical techniques. Comprehensive and far beyond the scope of these Lecture Notes.

1.2 The Climate System

1.2.1 Components of the Climate System

The climate system can be divided into five components (Fig. 1.1) which are introduced below. The overview mentions some important processes as examples:

1. *Atmosphere*: Gaseous part above the Earth's surface including traces amounts of other gaseous, liquid and solid substances. Weather, radiation balance, formation of clouds and precipitation, atmospheric flow, reservoir of natural and anthropogenic trace gases, transport of heat, water vapour, tracers, dust and aerosols.
2. *Hydrosphere*: All forms of water above and below the Earth's surface. This includes the whole ocean and the global water cycle after precipitation has reached the Earth's surface. Global distribution and changes of the inflow into the different ocean basins, transport of ocean water masses, transport of heat and

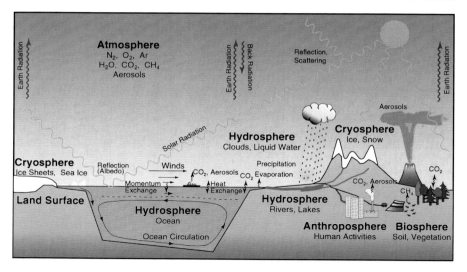

Fig. 1.1 The most important components and associated processes of the climate system on a global scale.

tracers in the ocean, exchange of water vapour and other gases between ocean and atmosphere, most important reservoir of carbon with fast turnover.
3. *Cryosphere*: All forms of ice in the climate system, including inland ice masses, ice shelves, sea ice, glaciers and permafrost. Long-term water reserves, changes of the radiation balance of the Earth surface, influence on the salinity in critical regions of the ocean.
4. *Land Surface*: Solid Earth. Position of the continents as a determining factor of the climatic zones and the ocean currents, changes in sea level, transformation of short-wave to long-wave radiation, reflectivity of the Earth's surface (sand different from rock, or other forms), reservoir of dust, transfer of momentum and energy.
5. *Biosphere*: Organic cover of the land masses (vegetation, soil) and marine organisms. Determines the exchange of carbon between the different reservoirs, and hence the concentration of CO_2 in the atmosphere, as well as the balances of many other gases, and therefore also the radiation budget. Influences the reflectivity of the surface, hence the radiation balance (e.g., tundra different from grassland), regulates the water vapour transfer soil-atmosphere, and via its roughness, the momentum exchange between the atmosphere and the ground.

A sixth component, which is particularly relevant for the assessment of future changes, is often treated as a distinct part of the climate system: the *anthroposphere* ($\alpha\nu\theta\rho o\pi o\sigma$ = human), consisting of the processes which are caused or altered by humans. The most important ones are the emission of substances which alter the radiation balance, and land use change (deforestation, desertification, degradation and transformation into constructed areas).

Most of the climate models treat processes and fluxes of the anthroposphere as an external forcing, i.e., the models are run by prescribing atmospheric concentrations and emissions of CO_2. Prescribed are also dust and sulphate emissions from volcanoes: for the past based on documented data and paleoclimatic information of volcanic eruptions, for the future they may be based on the statistics of such events.

A complete climate model contains physical descriptions of all five components mentioned above and takes into consideration their coupling. Some components may be described in a simplified form or even be prescribed.

Not all questions in climate sciences require a model comprising all components. It is part of the scientific work to select an appropriate model combination and complexity, so that *robust results* are produced for a specific science question.

Each climate system component operates on a range of characteristic temporal and spatial scales. The knowledge of these scales is necessary for a correct formulation of climate models. Table 1.1 summarizes some of relevant scales. Usually, the definition of processes to be represented in the model restricts the temporal and spatial resolution of the model's grid.

1.2.2 *Global Radiation Balance of the Climate System*

The Sun is the only relevant energy source for the climate system on a temporal scale of less than about 10^6 years. The different energy fluxes are shown in Fig. 1.2. Coming from the Sun, on average $341\,\text{W}\,\text{m}^{-2}$ reach the top of the atmosphere (this corresponds to about a quarter of the solar flux density, Solar Constant $S_0 = 1367\,\text{W}\,\text{m}^{-2}$), while barely half of this is available for heating of the Earth's surface. Major parts of the short-wave radiation are reflected by clouds or reflected directly on the Earth's surface itself and are absorbed by the atmosphere. Incoming radiation contrasts with surface long-wave outgoing radiation of around $396\,\text{W}\,\text{m}^{-2}$. Through convection and evaporation, the surface loses another $100\,\text{W}\,\text{m}^{-2}$, which would – if other important processes absent – result in a negative energy balance of the surface.

The natural *greenhouse effect*, caused by greenhouse gases such as H_2O, CO_2, CH_4, N_2O and further trace gases, is responsible for the infrared back-radiation of around $333\,\text{W}\,\text{m}^{-2}$. This results in an energy balance with a global mean surface temperature of about $14\,°\text{C}$.

1.3 Purpose and Limitations of Climate Modelling

Until around the early twentieth century, climate sciences were primarily concerned with the study of past climatic states. This was done by observation of the environment using mostly geological, geographical and botanical methods. By the end of the 1950s, important physical measurement methods were developed.

Table 1.1 Some examples of processes determining the climate with their characteristic time and spatial scales.

Component of the Climate System	Process	Characteristic time scale	Characteristic spatial scale
Atmosphere	Collision of droplets during cloud formation	10^{-6}–10^{-3} s	10^{-6} m
	Formation of convection cells	10^4–10^5 s	10^2–10^4 m
	Development of large-scale weather systems	10^4–10^5 s	10^6–10^7 m
	Persistence of pressure distributions	10^6 s	10^6–10^7 m
	Southern Oscillation	10^7 s	10^7 m
	Troposhere–stratosphere exchange	10^7–10^8 s	global
Hydrosphere	Gas exchange atmosphere–ocean	10^{-3}–10^6 s	10^{-6}–10^3 m
	Deep water formation	10^4–10^6 s	10^4–10^5 m
	Meso-scale oceanic gyres	10^6–10^7 s	10^4–10^5 m
	Propagation of Rossby waves	10^7 s	10^7 m
	El Niño	10^7–10^8 s	10^7 m
	Turnover of deep water	10^9–10^{10} s	global
Cryosphere	Formation of permafrost	10^7–10^9 s	1–10^6 m
	Formation of sea ice	10^7–10^8 s	1–10^6 m
	Formation of land ice masses	10^8–10^{11} s	10^2–10^7 m
Land surface	Changes in reflectivity	10^7–10^8 s	10^2 m – global
	Isostatic equilibration of the crust by covering ice masses	10^8–10^{11} s	10^6 m – global
Biosphere	Exchange of carbon with the atmosphere	10^4–10^8 s	10^{-3} m – global
	Transformation of vegetation zones	10^9–10^{10} s	10^2–10^7 m

The measurement of weak radioactivity of various isotopes was the basis for the dating of organic material and enabled the determination of flux rates in different environmental systems. The measurement of the stable isotopes in precipitation revealed a conspicuous temperature dependence. By analysing stable isotope ratios in permanently deposited water (i.e., polar ice) a natural "paleo-thermometer" was realised. The determination of the concentration of trace gases and other tracers in ice cores from Antarctica and Greenland made it possible, for the first time,

1.3 Purpose and Limitations of Climate Modelling

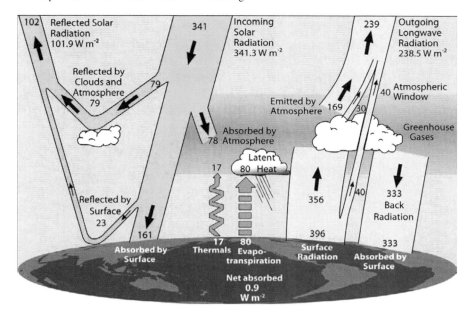

Fig. 1.2 Global energy fluxes from different sources which determine the radiation balance of the Earth. Figure from Trenberth et al. (2009).

to produce an accurate reconstruction of the chemical composition of the atmosphere. By exploring different *paleoclimatic archives*, which may be described as environmental systems that record and conserve physical quantities varying with time, an important step towards a quantitative science was taken. Such archives include ice cores from Greenland and Antarctica, ocean and lake sediments, tree rings, speleothems, and many more. This enabled the transition of climate science from the purely descriptive to a quantitative science providing numbers with units.

The increasingly detailed paleo-data require that hypotheses are quantitatively captured with regard to the mechanisms responsible for climate change. This is where *climate modelling* begins. Its goal is the understanding of the physical and chemical information and data retrieved from, among others, paleo-data. Such models permit a quantitative formulation and testing of hypotheses about the causes and mechanisms of past, and the magnitude and impact of future climate change.

Figure 1.3 visualizes the role of modelling in paleoclimate science in a schematic way. Climate change alters certain climate and environmental (C & E) parameters which then can be "read" using appropriate transfer functions. Even in this case, model formulation and application play a central role, but the term *climate modelling* is not applicable. Climate archives can only be made accessible to research by reliable measurement techniques. An experimental physicist produces climate data (e.g., the reconstruction of the atmospheric CO_2 concentration over the past 800,000 years). The modeler works on the development and application of models that yield model results within the framework process studies. The goal is the

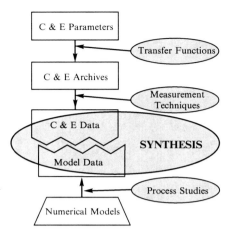

Fig. 1.3 The role of climate modelling in climate science. C & E stands for *climate and environmental*.

synthesis of model results and climate data, which is achieved when the underlying mechanisms and hypotheses are in quantitative agreement. Hence, the model yields *a quantitative interpretation of the evolution of climate*, based on the laws of physics and chemistry.

The evolution of the annual mean surface temperature averaged over the northern hemisphere over the course of the last 1,200 years is part of some of the most important climatic information in the debate on current climate change (Fig. 1.4). A central question, that has to be resolved by models, is whether the reconstructed warming – and what fraction of it – can be explained by the increase in atmospheric CO_2 and the resulting changes in the radiation budget. The modelling of the last 1,200 years of climate evolution necessitates an accurate knowledge of the different forcings to the radiation budget. The most important ones are the variations in solar radiation, the magnitude, location and duration of volcanic eruptions, the changes in land cover by deforestation and other activities and variations in concentration of climate-relevant atmospheric tracers. Besides sophisticated statistical methods, mainly climate models are able to answer these questions in a quantitative way.

The estimation of the *climate sensitivity*, that is the increase in the global mean temperature with a doubling of the atmospheric CO_2 concentration above the preindustrial level (from 280 to 560 ppm), is of major relevance in climate sciences. Models, that are employed to address this question, must be capable of simulating the natural climate variability as well as past climate changes.

A recent example is shown in Fig. 1.5. Here, the Bern2.5d model, a simplified climate model that describes the large-scale processes in the ocean and atmosphere, was used (Stocker et al. 1992; Knutti et al. 2002). The globally averaged warming, which is observed between 1860 and 2000 (grey band) can roughly be reproduced with different model simulations (lines). While the long-term trend is modeled in an acceptable way, single variations on a time scale of less than 10 years can only partly be captured. The uptake of heat by the ocean is only simulated in broad terms. The important deviations between 1970 and 1990 in ocean heat uptake may well

1.3 Purpose and Limitations of Climate Modelling

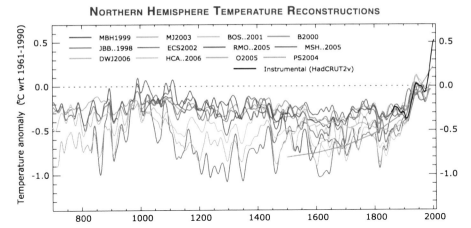

Fig. 1.4 Reconstructed annual temperatures in the northern hemisphere over the last 1,200 years, based on information from various paleoclimatic archives, primarily thickness of tree rings. A significant temperature increase over the last 100 years can be identified. On the other hand, several reconstructions show a very mild climate state around AD 1000 that clearly exceeds the ones at the beginning of the twentieth century. Figure from IPCC (2007), Technical Summary (Fig. TS.20, p. 55).

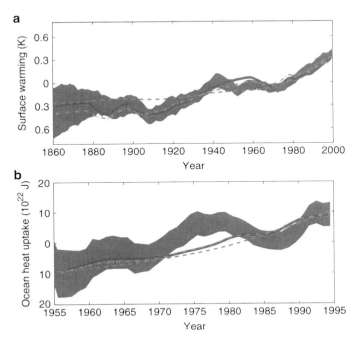

Fig. 1.5 Changes in global mean temperature since 1860 (**a**), and heat uptake in the ocean since 1955 (**b**). Grey bands for observations and lines for different model runs. Figure from Knutti et al. (2003).

be captured by particular simulations but, until today, have not been explained by climate models in a satisfactory way. However, this is a rare, but interesting example of a case in which a recent correction of the *observational* database has brought an improvement of the correspondence between experimental and computed data (Domingues et al. 2008).

As any mathematical model of natural systems, a climate model is a *simplification*. The degree of accepted simplification determines the complexity of the model and restricts the applicability of the model to certain questions. Hence, the complexity of a chosen model sets the limitations to its application. Determining these limitations requires considerable experience since no objective rules or guidelines exist. Especially for the development of climate models, particular care and a natural scepticism are needed: It is not desirable to implement and parametrise all processes without careful consideration of overall model consistency. The quality of a climate model is not judged by the mere number of processes considered, but rather by the quality of how chosen processes and their couplings are reproduced.

Of course, it is the duty of research and development to continuously increase the resolution and realism of climate models and this is happening at a fast pace. However, this rather quickly and regularly reaches the limits of computing resources particularly if long-term simulations (e.g., over 10^5 years or more) are performed. For this reason, intelligent simplifications and models of reduced complexity are required. This becomes manifest in the way how a *hierarchy of models* is used in current climate research. This will be discussed in Chap. 2.

1.4 Historical Development

Climate models emerged from models that were developed for weather prediction since around 1940. Modelling atmospheric processes and circulation is the cradle of climate model development. *Vilhelm Bjerknes* (1862–1951, Fig. 1.6) was the first

Fig. 1.6 Vilhelm Bjerknes (1862–1951), founder of dynamical meteorology.

1.4 Historical Development

Fig. 1.7 Lewis Fry Richardson (1881–1953) computed the first weather forecast.

to realize that weather prediction was a problem of mathematics and physics. Thus, conservation equations for mass, momentum and energy need to be formulated in order to calculate the dynamics of the atmospheric circulation. They are combined with an equation of state for an ideal gas. Hence, the atmosphere evolves in a deterministic way implying that consecutive states of the system are linked by physical laws.

Bjerknes assumed that a sufficiently accurate knowledge of the basic laws and the initial conditions were necessary and sufficient for a prediction. He therefore adopted the classical notion of predictability of nature, or determinism, from Laplace. Only later it will become apparent, most notably through the work of the late Edward Lorenz in 1963 (see Sect. 7.2), that the predictability of the evolution of a non-linear system, in this case the atmospheric circulation, is naturally limited. Bjerknes founded the "Bergen School" of meteorology and has produced ground-breaking contributions to the knowledge of cyclogenesis.

Lewis Fry Richardson (1881–1953, Fig. 1.7) was the first to formulate a numerically-based weather forecast. The calculations, which he conducted in 1917, were based on observational data from 12 vertical profiles of pressure and temperature at different stations across Europe, which – incidentally – were established by Bjerknes. These data served as initial conditions for the calculation. Richardson defined a calculation grid with a resolution of $3° \times 1.8°$ and five vertical layers across Europe. It consisted of 150 grid points, on which the pressure trends were calculated. Richardson made use of the so called primitive equations: the horizontal momentum conservation equations, the continuity equation (prescribing conservation of mass) and the ideal gas equation. The work load for the calculation of a 24-h forecast was enormous: It took three months. Only after the first computers were available in the 1940s, weather forecasts were feasible and were deployed as a tactical means by the end of the World War II. Richardson's first computations were a significant achievement of principle value but did not provide reliable predictions. The prediction for the change in surface pressure over 6 h yielded a value of 145 hPa. Not even in the center of a low-pressure system such a fast drop in pressure can be observed. Nevertheless, Richardson published his result in the famous book *Weather Prediction by Numerical Process* (Richardson 2007). The problem was

Fig. 1.8 Stationary Rossby waves in a rotating tank (http://www.ocean. washington.edu/research/ gfd).

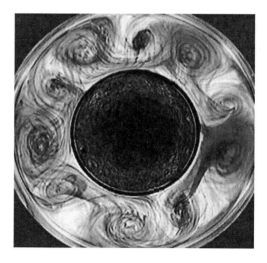

that the initial conditions, in this case the data for the surface pressure, contained small errors that multiplied during the numerical procedure and led to strong trends in pressure. A calculation based on the same data but filtered at the beginning by adjusting unnaturally strong pressure gradients, led to a plausible prediction with Richardson's algorithms (3.2 hPa/6 h).

This points to the fact that initial conditions, or the *initialization* of weather and climate models, is a central problem of which the modeler must always be aware. Not only the initial conditions, but also the formulation of conservation equations is crucial. Even the most accurate initial data would have led to instability using the equations of Richardson, because they contained physical processes (gravitational waves), that destabilize the solution and make a long-term prediction impossible.

Carl-Gustav Rossby (1898–1957) achieved a break-through by realizing that the conservation of *vorticity* was a more robust constraint than that of momentum. This approach is suitable for the system of the rotating Earth, because the Coriolis effect can be implemented in a natural way. Planetary waves (Rossby waves) appear in rotating fluids (Fig. 1.8) such as the atmosphere and the ocean. Atmosphere and ocean respond to disturbances (temperature anomalies, onset of deep water formation, etc.) with the propagation of Rossby waves that cause currents which then are able to modify the background state. Rossby waves are fundamental for the understanding of weather systems in the atmosphere and the large-scale circulation in the ocean. Interesting further information is provided at http://www.ocean. washington.edu/research/gfd including many descriptions of table-top experiments in geophysical fluid dynamics.

In the 1940s and 1950s the first computer (ENIAC, Electronic Numerical Integrator and Computer) was deployed in Princeton for the US Army. The first project was the prediction of a storm surge at the American East Coast. In 1955, the first long-term integrations of a simplified atmospheric circulation model were realized by

1.4 Historical Development

Fig. 1.9 Edward Lorenz (1926–2008), the inventor of chaos theory.

Norman Phillips (Phillips 1956). This marked the beginning of *general circulation models* which would solve the complete equations of atmospheric flow.

Besides the numerically complex problems, theoretical studies on the fundamentals of the dynamic behaviour of the atmosphere and the ocean were advanced. The conservation of momentum and vorticity in a rotating fluid implies non-linear terms in the equation system. They result from advection of momentum in a flow (terms of the form $u\,\partial u/\partial x$, etc.). In addition, in a rotating frame such as the Earth, the Coriolis force causes a coupling of the components of the horizontal movements. Non-linearities are responsible for the finite predictability of such flow as *Edward Lorenz* (1926–2008, Fig. 1.9) has found in 1963. In his landmark paper *Deterministic non-periodic flow* (Lorenz 1963) he describes how the patterns of large-scale flow can lead to chaotic behaviour (see Sect. 7.2).

This pioneering paper set the basis for a entirely new scientific domain: *Chaos Theory*. Although, the evolution of a classical system can be calculated in a deterministic way at all times (by solving partial differential equations), the system loses its predictability after a finite time. Smallest differences in the initial conditions may result in totally different states already after a short time. A scaling of the final state as a function of the initial states is not possible anymore. This finding is well known as the "butterfly effect". An excellent book with many reminiscences and mathematical examples is *The Essence of Chaos* by Edward Lorenz (Lorenz 1996).

In the mid 1960s, almost 20 years after the development of the first models for the circulation in the atmosphere, three-dimensional ocean models were formulated (Bryan and Cox 1967).

Syukuro Manabe (Fig. 1.10) found that for climate research atmospheric and oceanic models need to be combined. This is achieved by dynamically coupling the two components. The first coupled model was developed in the late 1960s by Suki Manabe and colleagues (Manabe and Bryan 1969). A particular difficulty was the completely different time scales for the atmosphere and the ocean (see Table 1.1). A notorious problem was that the required heat and water fluxes from the atmosphere and the ocean, which yield climatologies that are coherent with observations, were not compatible. This necessitated the introduction of a non-physical *flux correction*, which was used in most of the models over almost 30 years.

This topic will be discussed in Chap. 8.6. The problem could only be resolved in the last decade thanks to a higher resolution of the models – generally a resolution

Fig. 1.10 Suki Manabe, pioneer of coupled climate modelling, during a reception in Tokyo in 2004.

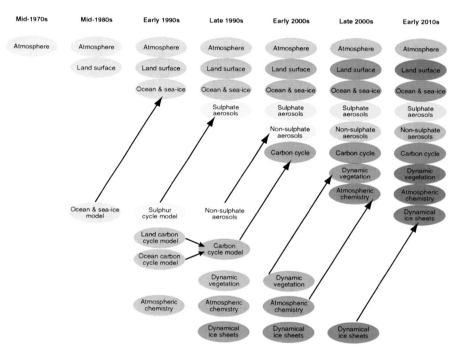

Fig. 1.11 Chronology of climate model development. The implementation of new components (carbon cycle, vegetation and atmospheric chemistry) leads to an increased complexity as well as to an increase in required computational resources. Yet it is a necessary development when the interaction of the different processes needs to be simulated quantitatively. Figure modified from IPCC (2001), Technical Summary (Box 3, Fig. 1, p. 48).

of at least $2° \times 2°$ is required, as well as due to improved parameterisations of not explicitly resolved processes.

Since the early 1990s, significant improvements were achieved by incorporating further climate system components (Fig. 1.11). Climate models have become more complete. The carbon cycle, dynamical formulations of vegetation types, the chemistry of the atmosphere and ice sheets, belong to components that are currently implemented into existing physical circulation models. In consequence, climate modelling has become an *interdisciplinary* science.

Besides ever more detailed models, also simplified climate models are being developed. They permit the study of basic problems of climate sciences in an efficient way. The development and application of climate models of reduced complexity (often called EMICs, *Earth System Models of Intermediate Complexity*) have made important contributions to the understanding of the climate system, in particular in the quantitative interpretation of paleoclimatic reconstructions and ensemble simulations of future climate change.

1.5 Some Current Examples in Climate Modelling

1.5.1 Simulation of the Twentieth Century to Quantify the Link Between Increases in Atmospheric CO_2 Concentrations and Changes in Temperature

Given that the most important driving factors of the radiation balance are known, the effect of increasing CO_2 concentrations on the annual mean atmospheric temperature can be estimated. Figure 1.12 presents the results of simulations with climate models using slightly different initial conditions (so called ensemble simulations). The averaged temperatures of the model runs are compared with observations during the twentieth century (*bold lines*). If the models consider all driving factors: change in the solar "constant", volcanic eruptions, atmosphere–ocean interactions, changes in the concentration of CO_2, other greenhouse gases as well as sulphate aerosols, agreement of the simulations with the observational records is found (red bands). In case the anthropogenic driving factors are held constant, a systematic deviation of all model simulations from the data appears from 1970 onwards (blue bands). This finding is valid globally, as well as averaged over continental scales.

This leads to a clear statement:

> Most of the observed increase in globally averaged temperature since the mid-twentieth century is *very likely* due to the observed increase in anthropogenic greenhouse gases.

which was made in the Fourth Assessment Report of the Intergovernmental Panel on Climate Change, IPCC (2007).

1.5.2 Decrease in Arctic Sea Ice Cover Since Around 1960

The decrease in the Arctic ice cover is documented by direct observations as well as by remote sensing. Since around 1960, the decrease in total area has accelerated (Fig. 1.13). Evidence from submarine missions also points to a drastic decrease in the thickness of sea ice. A similar development is visible in all coupled climate models which were used for the Fourth Assessment Report of the Intergovernmental Panel on Climate Change, IPCC (2007). The models indicate an accelerated decrease in the extent of Arctic sea ice since around 1960. The simulations assume an increase in CO_2 from 1990 onwards, prescribed according to emission scenario A1B (rapid economic growth, balanced emphasis on all energy sources, see Nakicenovic et al. 2000). While observations and model simulations agree with

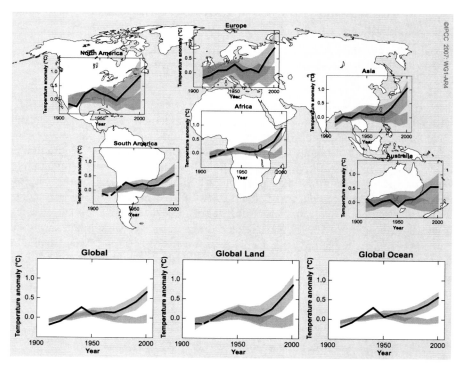

Fig. 1.12 Global and continental evolution of the temperature since 1900, based on measurements (*bold line*) and ensemble simulations with coupled climate models (*bands*). Only simulations with a complete forcing which includes changes in greenhouse gases, aerosols, observed volcanic eruptions and variable solar radiation, show reasonable agreement with the observations over the entire twentieth century (*red bands*). In case the effect of anthropogenic forcings (greenhouse gases, aerosols) on the radiative balance is not taken into account, the global and continental-scale increase in temperature cannot be simulated (*blue bands*). Figure from IPCC (2007), Summary for Policymakers (Fig. SPM.4, p. 11).

1.5 Some Current Examples in Climate Modelling

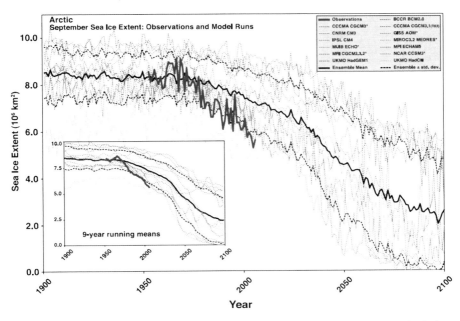

Fig. 1.13 Changes in sea ice cover in the Arctic from 1900 according to multi-model simulations for IPCC (*fine curves*) and observations shown in the *red curve*. Figure from Stroeve et al. (2007).

negative trends of Arctic sea ice cover, the observed trend since 1960 is significantly stronger. While this points to a general shortcoming of current models to predict this important variable, it also issues a strong warning regarding the development of Arctic sea ice cover in the next few decades. In fact, we cannot exclude a sea ice-free Arctic in late summer by the year 2030.

1.5.3 Summer Temperatures in Europe Towards the End of the Twenty-First Century

The question how an increase in global mean temperatures will affect the climate in Europe can still only roughly be answered by a few climate models with regional resolution (Fig. 1.14). The high resolution (56 km) requires enormous computational resources and only so called *time slices* can be calculated. The simulation with a regional climate model shows a significant increase in summer temperatures in Europe between 2071 and 2100 (Schär et al. 2004). The warming is accentuated at high altitudes due to the positive snow-albedo feedback and in the Mediterranean area due to the positive feedback of soil moisture. Besides a strong warming by the end of the twenty-first century, every second or third summer then will be equally hot or hotter than the extreme summer of 2003, an extreme event which had not occurred in the last 500 years.

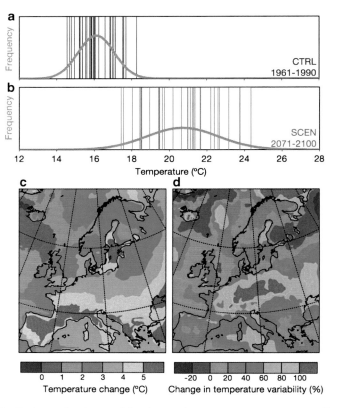

Fig. 1.14 Distribution and estimate of the changes in summer temperatures over Europe in the years 2070–2100, calculated with a regional climate model (Schär et al. 2004). Panels (**a**) and (**b**): Distribution of summer temperatures for 30 years in the 20th century (CTRL) and 30 years at the end of the 21st century (SCEN). Panels (**c**) and (**d**): Temperature change and change in temperature variability between CTRL and SCEN.

A single simulation, however, is not yet a reliable description of the expected warming. Therefore, ensemble simulations with individual models and the aggregation of such into multi-model ensembles have become the standard. Uncertain quantities such as the climate projections or the influence of clouds must be examined systematically. Future climate projections will be associated with estimates of probability which can be derived from *ensemble simulations*. This approach has already been used for the Fourth Assessment Report of the Intergovernmental Panel on Climate Change, IPCC (2007).

1.5.4 CO_2 Emissions Permitted for Prescribed Atmospheric Concentration Paths

How much greenhouse gases, for example CO_2, may be emitted each year without exceeding the tolerated concentrations of these gases? The answer to this question

1.5 Some Current Examples in Climate Modelling

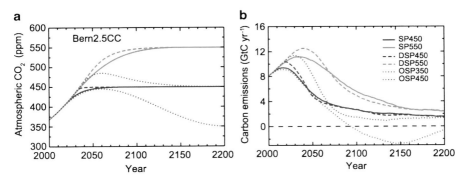

Fig. 1.15 Projected allowable carbon emissions leading to stabilization of atmospheric CO_2 at given stabilization levels for the Bern2.5CC EMIC for different pathways leading to stabilization. (**a**) Assumed trajectories of CO_2 concentrations in the SP, OSP, and DSP profiles. (**b**) Implied carbon emissions as projected with the Bern2.5CC EMIC. Profiles with the delayed turning point in the atmospheric CO_2 increase (DSP) or atmospheric CO_2 overshoot (OSP) are compared to the standard SP profile. Thirty-one-year running averages are applied to the results. Figure from Plattner et al. (2008).

can only be given with the aid of climate models that include representations of biogeochemical cycles, in particular the carbon cycle. The exchange with the ocean and the role of the terrestrial and marine biosphere have to be considered with suitable sub-models and parameterisations.

Figure 1.15 shows an example calculated at the Division of Climate and Environmental Physics, University of Bern, Switzerland, with a simplified climate model. The long-term stabilisation of CO_2 concentrations can only be achieved by strongly reduced and ultimately vanishing emissions of CO_2. This would require a complete replacement of fossil fuels. In 1998, the emissions of all fossil energy sources (cement production included) was around $6.6\,\text{GtC}\,\text{yr}^{-1}$ ($1\,\text{GtC}\,\text{yr}^{-1} = 1$ gigaton carbon $\text{yr}^{-1} = 10^{12}\,\text{kg}\,\text{C}\,\text{yr}^{-1}$); ten years later it was exceeding $8\,\text{GtC}\,\text{yr}^{-1}$. The computations show that after a permitted maximum in 2030, the emissions need to decrease drastically (globally around 1% per year). Such model simulations are of crucial significance to global political decisions related to international treaties such as the Kyoto-Protocol and its successors.

1.5.5 Prediction of the Weak El Niño of 2002/2003

The irregular warming of waters in the tropical Eastern Pacific, known as the *El Niño-Southern Oscillation* (ENSO) phenomenon, strongly affects the tropical climate and in particular the water cycle. The formation of atmospheric pressure and temperature anomalies also causes deviations from the usual climate around the globe (*teleconnections*).

These changes, which may last some months up to around 1.5 years, cause severe economic damage. Due to the various teleconnections, some regions may exist which are affected by El Niño in a positive way (e.g., by increased precipitation in vegetation regions, where water is normally the limiting factor). However, the strong El Niño of 1997–1998 is estimated to have caused net economic damage (gains and losses, depending on the region) in the USA of around 25 billion US$. Therefore, a reliable prediction of El Niño is of highest economic and societal significance.

For the first time, the ENSO event of 1997–1998 could be predicted already 6 months in advance. This time span allowed the affected regions to take precautions and to adapt to the expected climatic consequences (droughts, floods, poor harvest, increased prevalence of Malaria by unusually high temperatures, etc.). This success was enabled by intensive research in the theory of the coupling between ocean and atmosphere in the tropics, model development and set-up of a dense observation net in the tropical Pacific (in situ and via remote sensing) since the early 1980s (TOGA Program).

Figure 1.16 shows the prediction of the evolving ENSO 2002/2003, as it was available in August 2002. A moderate increase in SST (*sea surface temperature*) in the tropical Eastern Pacific (right) was expected. It is important to note that the

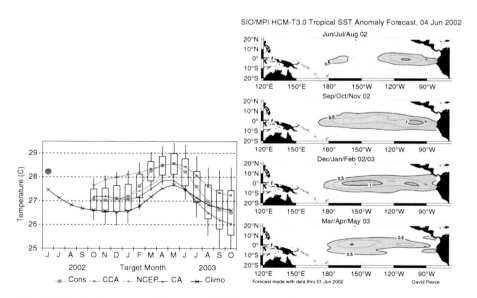

Fig. 1.16 Left: Temperature in the tropical Eastern Pacific, based on several models, initialized with data until August 2002. Figure from http://www.cpc.ncep.noaa.gov/products/predictions/90day/SSTs, National Oceanic and Atmospheric Administration (NOAA), National Weather Service (NWS), Climate Prediction Center (CPC). **Right**: Distribution of SST anomalies (sea surface temperature) from summer 2002 to spring 2003 based on a global coupled climate model. Figure from http://grads.iges.org/ellfb/Jun02/pierce/fig1.gif, Institute of Global Environment and Society (IGES).

single models differ in their quantitative prediction. Hence, the prediction bears an uncertainty, analogous to the daily weather forecast in which the occurrence of rain is also given with a probability.

1.6 Conclusions

Climate models are simplified descriptions of complex processes within the climate system. They are used for the quantitative testing of hypotheses regarding the mechanisms of climate change, as well as for the interpretation of instrumental data from paleo-data from various archives. Climate models are essential for the operational prediction of the economically important ENSO-phenomenon and other climate modes. A further important motivation for the development and application of climate models remains the aim to assess future climate change.

Research developing and using climate models has become interdisciplinary and comprises domains of physics (thermodynamics, fluid dynamics, atmospheric physics, oceanography), chemistry (organic, inorganic and surface chemistry, reaction kinetics, geochemistry, cycles of carbon, nitrogen, etc.) and biology (vegetation dynamics, ecology).

By the end of the 1960s, simple climate models (energy balance models) were developed in order to examine planned climate modifications (Budyko 1969). The idea was to strongly reduce the snow cover by a large-scale distribution of ash and therefore cause a warming of Siberia in order to access new agricultural lands ("*geo-engineering*"). In the meantime, we have become aware that humans alter the climate inadvertently by continuous emissions of CO_2 and other greenhouse gases. The increase in atmospheric CO_2 concentrations (Fig. 1.17) testifies to this fact with great precision. This time series has become a corner stone in global change research.

Figure 1.17 also provides evidence of life on planet Earth and shows its global signature. The seasonal fluctuations in CO_2 are the result of the "breathing" of the biosphere (vegetation and soils). During spring in the Northern Hemisphere, carbon is taken up and is released in winter through respiration. Additionally, the interannual variability of CO_2 is visible, which is caused by the warming and cooling of large parts of the ocean, for example during ENSO events or volcanic eruptions.

Today, CO_2 concentrations are 28% higher than ever before in the last 800,000 years (Lüthi et al. 2008). This important fact has been derived from several decades of research on ice cores from Greenland and Antarctica. Ice contains bubbles in which air is enclosed. The enclosure process occurs at the firn-ice transition in a depth of about 80–100 m on the two polar ice sheets of Greenland and Antarctica. Ice cores are therefore natural archives which preserve information on the content and composition of the atmospheric air in the past.

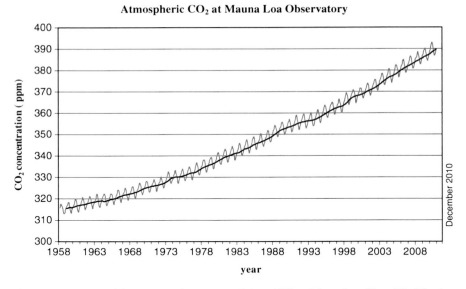

Fig. 1.17 Increase in CO_2 concentration, measured since 1958 on Mauna Loa (Hawai'i). CO_2-data from http://www.esrl.noaa.gov/gmd/ccgg/ (Dr. Pieter Tans, NOAA/ESRL).

Figure 1.18 shows a compilation of such measurements of CO_2 and an estimate of local temperature based on the concentrations of the stable isotopes in ice. At the far right side the increase in CO_2 during the last 250 years is added to the graph. The CO_2 measurements in the older half of this time series were performed at the University of Bern (Siegenthaler et al. 2005; Lüthi et al. 2008). This demonstrates not only the unprecedented concentrations of CO_2 over the last 800,000 years, but also the rate of increase of CO_2 which is estimated to be 100 times faster than ever during the last 20,000 years.

Regarding the ongoing changes in the composition of the atmosphere and land-use, and the climate change induced by them, the global community has defined a remarkable goal in Article 2 of the UN Framework Convention on Climate Change (UNFCCC, 1992, http://unfccc.int):

> **Article 2:** The ultimate objective of this Convention and any related legal instruments that the Conference of the Parties may adopt is to achieve, in accordance with the relevant provisions of the Convention, stabilization of greenhouse gas concentrations in the atmosphere at a level that would prevent dangerous anthropogenic interference with the climate system. Such a level should be achieved within a time-frame sufficient to allow ecosystems to adapt naturally to climate change, to ensure that food production is not threatened and to enable economic development to proceed in a sustainable manner.

1.6 Conclusions

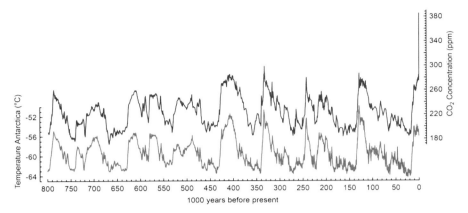

Fig. 1.18 Evolution of the atmospheric CO_2 concentration (*blue*) and Antarctic temperature (*red*) over the past 800,000 years from measurements on several ice cores from Antarctica (Petit et al. 1999; Siegenthaler et al. 2005; Jouzel et al. 2007; Lüthi et al. 2008). Direct measurements of CO_2 in the atmosphere since 1958 are added.

In the light of this global change, geo-engineering has experienced a recent revival, and climate models are now used to quantify the consequences of human climate modification (Robock et al. 2008). However, no proposal so far has convincingly shown that geo-engineering is able to reduce global warming without other, undesired side-effects and therefore, many reasons can be brought forward, not least moral ones, to reject the option of geo-engineering (Robock 2008).

Chapter 2
Model Hierarchy and Simplified Climate Models

2.1 Hierarchy of Physical Climate Models

There is no *best* climate model! Different models have different advantages which may be due to their complexity or the form of their implemented parameterisations. Table 2.1 gives an (incomplete) overview of the hierarchy of models used for climate simulations. They are ordered according to their spatial dimensions. Only model types are listed but each type may be formulated in different ways. For instance different resolutions are used, different grid structures, parameters and parameterisations are chosen in a different way, etc. There are, for example, more than a dozen different ocean circulation models, all of which basically solve the same conservation equations. For model development and progress the various *Modelling Intercomparison Projects* provide important insight: AMIP (*Atmospheric Modelling Intercomparison Project*), OMIP (*Ocean...*), OCMIP (*Ocean Carbon-cycle...*), CMIP (*Coupled...*), PMIP (*Paleo...*), C^{4}MIP (*Coupled Climate-Carbon Cycle Modelling Intercomparison Project*), etc.

In order to tackle problems across the board in climate dynamics, a model hierarchy is required. An example is the investigation of the climate at the time of the *Last Glacial Maximum* some 21,000 years ago. Simplified models of the type shown in the grey shaded area of Table 2.1 permit a systematic examination of the parameter space: which driving factors (radiation, precipitation) are important for simulating, for example, the water mass distribution in the ocean, which parameters and processes produce a significant cooling of the tropics, etc.

Models of spatial dimension 0 or 1 help us illustrate some fundamental concepts in climate dynamics. Clever formulations of these 0-dimensional models are, under given circumstances, very useful for scenario or ensemble calculations. An EBM point model will be presented in Sect. 2.2.

So called *Saltzman Models* are globally averaged models which simulate some time dependent, large-scale variables (e.g., global mean temperature, ice volume, CO_2 content, etc.) and form a non-linear, dynamical system. These models can be derived from the basic equations in a rigorous way (Saltzman 2001). They are a radical alternative to the classic approach in climate modelling and yield some interesting hypotheses. For example, the question regarding the origin of the transition

Table 2.1 Hierarchy of coupled models for the ocean and the atmosphere with some examples, ordered according to the number of spatial dimensions considered. The direction of dimensions is specified in brackets (lat = latitude, long = longitude, z = vertical); 2.5d corresponds to several two-dimensional ocean basins linked in the Southern Ocean; EBM stands for *energy balance model*; QG is the abbreviation for quasi-geostrophic, AGCM (*atmospheric general circulation model*), OGCM (*ocean general circulation model*), SST (*sea surface temperature*). Some example models and their authors are given in italics, the grey shaded area contains climate models of reduced complexity, also called *Earth System Models of Intermediate Complexity* (EMICs), which permit integrations over very long periods (several $10^3 - 10^6$ years) or large ensembles. The table is not completely full because some combinations are not meaningful.

Dimension		Ocean			
		0	1	2	3
Atmosphere	0	EBM point models; Bipolar seesaw *Stocker & Johnsen*; Dynamical systems: *Saltzman models*; pulse-response-models *Siegenthaler/Joos*; Neural Networks *Knutti et al. 2002*	Ekman models (z); global mixing (z) *Munk*; Advection-diffusion model (z): *HILDA Bern Wigley-Raper*	thermohaline models (lat/z): *Stommel, Marotzke*; wind-driven flow (lat/long): *Stommel, Munk*; Deep ocean (lat/long): *Stommel, Pedlosky*	OGCM
	1	EBM (lat) *Budyko, Sellers*; radiative-convective model (z) *Manabe*	–	ocean (lat/z) + EBM (lat): *Bern2.5d model Stocker, Wright, Mysak*	–
	2	EBM (lat/long) *North and Crowley*	statistical dynamical atm. (lat/z) + diffus. ocean (z): *MIT model*	ocean (lat/z) + statistical dynamical atm. (lat/long): *Climber2*	OGCM + EBM (lat/long): *UVic model, Bern3D*; OGCM + QG atm. model: *ECBILT-CLIO*
	3	AGCM + SST	AGCM + mixed layer	AGCM + slab ocean	AOGCM *CCSM3, HadGem, etc.*

from a 40,000- to a 100,000-year periodicity of the glacial cycles about 10^6 years ago can be addressed with such model formulations.

Pulse response models are efficient substitute models for particular quantities which are simulated in a more comprehensive and expensive way by three-dimensional models. They require a linear behaviour of the simulated processes which at first has to be verified by a more complex model. The response of a complex model to any disturbance (for example the warming caused by an increase in atmospheric CO_2) can be regarded as a temporal integral of elementary responses of a complex model to a pulse-like perturbation (δ-function). These models are, e.g., successfully applied to the calculation of CO_2 uptake by the ocean or for the global warming as an input for vegetation models. Thanks to their simplicity, they permit extended scenario calculations.

A not yet common but promising method is the application of *neural networks* with which substitutes for complex climate models can be built. In contrast to *pulse response models* processes that are non-linear or include several equilibria can be substituted. A limiting factor is the fact that neural networks need to be trained with simulations of the model to be substituted. Since such "training sets" require information, a certain amount of computational effort is necessary. Once the network is trained, the calculation of ensembles can be realized very efficiently. This method was employed using a simplified model (Knutti et al. 2003).

Energy balance models (EBM) belong to the earliest simplified climate models that were used for the quantitative assessment of climate change. An example shall be discussed later in Sects. 2.2 and 4.3.

Advection-diffusion models describe, e.g., the vertical mixing in the ocean on a global scale in a summarized form. They provide insight into some aspects of the carbon cycle (e.g. Siegenthaler and Joos 1992); they are applied for questions concerning past changes in atmospheric CO_2 (last 10,000 years) as well as for the assessment of emission scenarios for future climate change.

Models of the category (0/2) are theoretical models of physical oceanography, but some of them are used as ocean components in simplified climate models. The class of climate models of reduced complexity (*Earth System Models of Intermediate Complexity*) is shaded in grey in Table 2.1. Long-term simulations, particularly important for paleoclimate dynamics, are based on such models.

The Division of Climate and Environmental Physics at the University of Bern, Switzerland, has developed and applied such models since 1993. The model concept and the extremely simplified geometry are shown in Fig. 2.1. Although only very few atmospheric and oceanic processes are considered, and the number of parameterisations is kept at a minimum, these models are fairly consistent with observations on large spatial scales ($>10^6$ m). For example, the meridional distribution of air temperature or the distribution of water masses in the three ocean basins compare well with observational estimates. These models were successfully employed in various ways in order to simulate quantitatively past climate change as, for example, found in Greenland ice cores. Even some basic aspects of biogeochemical cycles were implemented which permitted the direct comparison of model results with ice core measurements of CO_2 and other greenhouse gases (Marchal et al. 1999).

These models were also used to assess the stability of the oceanic circulation in the Atlantic under a global warming scenario. The models showed that the stability of the circulation not only depends on the absolute amount of warming, but also on the rate of warming (Stocker and Schmittner 1997). Later, this fundamental

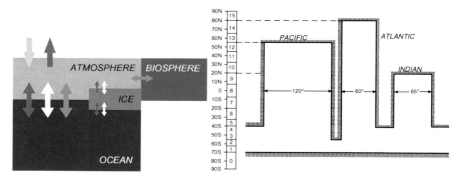

Fig. 2.1 Concept and geometry of the Bern2.5d model, one of the first climate models of reduced complexity (Stocker et al. 1992). Ocean currents are averaged zonally and are simulated by three basins, connected in the south (category 1/2). Thanks to the strongly simplified depiction of the climate system, simulations spanning over 10^6 years are possible.

finding was confirmed by three-dimensional AOGCMs (*Atmosphere/Ocean General Circulation Models*). This is a good example for how new and relevant climate mechanisms are found and explored with models of reduced complexity. Of course, such results then need to be verified or falsified by more comprehensive models. The implementation of suitable biogeochemical components permits the examination of the interaction of the carbon cycle with the ocean over the course of the next 1,000 years (Joos et al. 1999; Plattner et al. 2008; see also Fig. 1.15). This is of significance for the question of a possible *run-away greenhouse effect* as a result of an anthropogenic increase in atmospheric CO_2. In the future, such models (e.g., the MIT model in category 2/1) may be coupled to macro-economic models, which assess the economic effects of climate change and mitigation options.

The latest developments at the Division of Climate and Environmental Physics, University of Bern, are devoted to models of category 3/2, where the ocean is three-dimensional, but coarsely resolved. This model type can be combined with biogeochemical modules and represents an important novel instrument in paleoclimate research (Müller et al. 2006; Ritz et al. 2008, 2011).

Comprehensive climate models consist of a three-dimensional formulation for the atmosphere (AGCM, *Atmospheric General Circulation Model*) as well as for the ocean (OGCM, *Ocean General Circulation Model*). The grid structures are shown schematically in Fig. 2.2. The coupling of the two, often given in differently formulated grids, is dynamic, meaning that ideally, at each time step, momentum, heat and water, and other tracers, are exchanged. For sufficiently good models, this is possible in a consistent way. Otherwise *flux corrections* have to be implemented in order to stabilize the simulated climate.

AGCMs, OGCMs and AOGCMs are classified in the highest levels of the model hierarchy shown in Table 2.1. They are extremely demanding with regard to their development, maintenance, computer time and storage and, finally, the analysis of

2.1 Hierarchy of Physical Climate Models

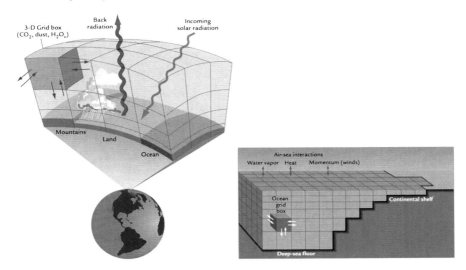

Fig. 2.2 Schematic illustration of model grids in three-dimensional AGCMs and OGCMs. The resolution of a coupled climate model is typically set at 4° × 4° to 2° × 2° and 20–40 vertical layers. Today, for single components, resolutions of up to 0.1° are applied. In this case, the calculation is restricted to either limited regions or an extremely short time of integration, hence not yet applicable for global climate studies. Figures from Ruddiman (2007).

results. Although such models are already run on personal computers or clusters, for their integration period quite strong limitations exist. A simulation of a hundred years is already a large project! These models contain a large number of parameterisations. They are being developed at various centers globally (Hadley Centre, UK; MPI Hamburg, DE; NCAR, USA; NASA-GISS, USA and many others).

The agreement of the latest climate models with observations is already respectable even for complex quantities such as water vapour (Fig. 2.3). The atmosphere consists of a rich structure of regions that are very dry (between 20° and 45° in latitude) and regions that are very humid with over 90% of humidity (tropics and 50°–65° in latitude). Models with highest resolution (around 1° × 1°) are capable to simulate even very strong gradients, similar to what is observed by satellites.

Model intercomparisons still reveal large deviations among the models (Fig. 2.4). Generally, heat, as opposed to water vapour, shows the smallest uncertainties and the different models are more consistent. Model differences are largest in polar regions where one has to deal with numerical challenges due to the convergence of the model grid and temperature differences of over 20°C occur between models. In the tropics, where cloud formation significantly affects surface temperatures, model differences are also increased.

Precipitation belongs to the most difficult components in climate modelling. Hence, deviations between the different models are large for all variables of the water cycle. Figure 2.5 shows the simulated annual-mean, zonally averaged

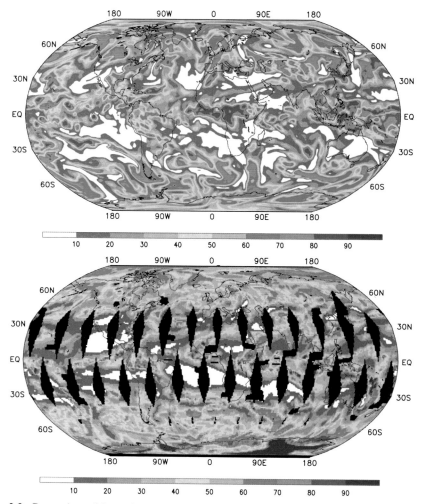

Fig. 2.3 Comparison of the performance of a climate model considering atmospheric water vapour content at 400 hPa (around 7 km height), given as relative humidity (in %) on a day in May. **Above**: model simulation with the model of MPI Hamburg at high resolution (T106, Wild (2000)). **Below**: Mean relative humidity between 250 and 600 hPa, based on satellite data (SSM/T-2), while uncertainties outside 30°S and 30°N are larger. Dry regions can be identified as white areas. Figure from IPCC (2001), Chap. 7 (Fig. 7.1a and c, p. 424).

precipitation amount. In the model evaluation preformed in IPCC (2001) deviations were large in regions with high precipitation (tropics, mid-latitudes) and may have differed between extreme model versions by as much as a factor of 2. However, the model evaluation performed in IPCC (2007) has shown some very considerable progress in the reliability and realism of the simulation of the global water cycle.

2.1 Hierarchy of Physical Climate Models

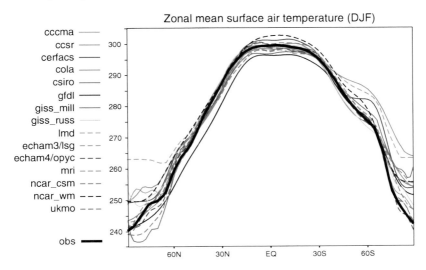

Fig. 2.4 December–February temperatures in Kelvin, zonally averaged for 15 global AOGCMs and comparison with the observed climatology (*black line*). Figure from IPCC (2001), Chap. 8 (Fig. 8.2, p. 480).

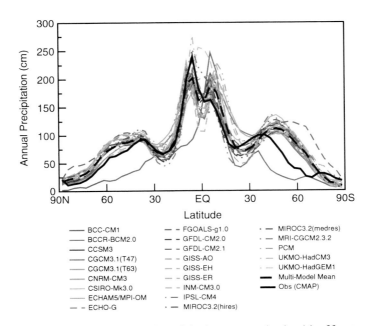

Fig. 2.5 Annual-mean, zonally-averaged precipitation amount simulated by 23 comprehensive coupled climate models used in IPCC (2007). Observed estimates are shown by the *black bold line* and they are compared with the multi-model mean (*black dashed*). Figure from IPCC (2007), Chap. 8, Supplementary Materials (Fig. S8.10, p. SM.8-49).

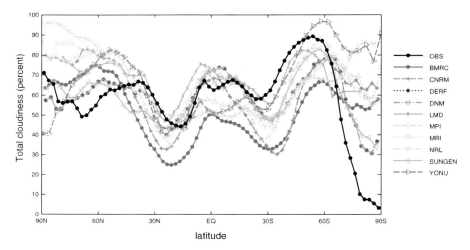

Fig. 2.6 December–February cloud cover in percent, zonally averaged for 10 AGCMs and its comparison to the observed climatology (*black line*). Figure from IPCC (2001), Chap. 8 (Fig. 8.7, p. 485).

Deviations are reduced, but still systematic differences to the observations persist, e.g., in the southern hemisphere (Fig. 2.5). For this reason, climate projections regarding rain and associated extreme events are still uncertain.

Another important quantity is the distribution of cloud cover since it strongly affects the water vapour feedback effect. A model intercomparison is given in Fig. 2.6. It shows the winter cloud cover (December–February), as simulated by 10 AGCMs. The deviations between them polewards of 60°, and in some cases from the observations, are considerable. Such comparisons reveal the current limitations of climate models and point to the necessary improvements (grid resolution, parameterisations, etc.).

A complete overview of the characteristics of different climate models and their comparison is given in IPCC (2001), Chap. 8, and can be viewed under http://www.grida.no/climate/ipcc_tar/wg1/308.htm, as well as more recent results from IPCC (2007), Chap. 8, Supplementary Materials.

Model development has made significant progress in the past decade. Simulated large-scale precipitation patterns of multi-model means now compare quite well with observations. The averaging process smoothes out deviations of individual models and produces a distribution of precipitation which compares well with observations (Fig. 2.7). This is why, for the first time, a credible large-scale projection of future changes in precipitation could be provided in IPCC (2007).

In preparation of the Fourth Assessment Report of IPCC, all major modelling centers delivered standardized model simulations for the twentieth and twenty-first century for a reduced set of emissions scenarios. The results are centrally stored at Lawrence Livermore National Laboratories and made available to the science community through the Program for Climate Model Diagnosis and Intercomparison

2.1 Hierarchy of Physical Climate Models 33

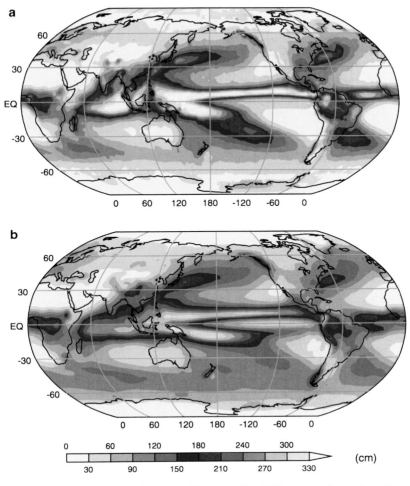

Fig. 2.7 Annual mean precipitation (cm), (**a**) observed and (**b**) simulated, based on the multi-model mean for the period 1980 to 1999. Grey regions in (**a**) indicate missing observations. Figure from IPCC (2007), Chap. 8 (Fig. 8.5, p. 612).

(PCMDI). It can be accessed through www-pcmdi.llnl.gov/ipcc/about_ipcc.php and has been used extensively during the past few years for model comparison and the investigation of climate model response to increasing greenhouse gases. A new effort, coordinated within the framework of the Coupled Climate Modelling Intercomparison Project, Phase 5, CMIP5, is currently ongoing. The IPCC Fifth Assessment Report will heavily draw from these results.

2.2 Point Model of the Radiation Balance

For illustrative purposes we consider first the simplest of all possible climate models with 0 dimensions. A single conservation equation for the globally integrated heat content is formulated (see Table 2.1, 0/0). Even though the model is not of great importance, it is instructive in various aspects. Using this simple example we will show how solutions of climate models fundamentally depend on the exact choice of parameterisations.

In this example which can be solved analytically in simple cases we can also discuss basic numerical schemes which are employed in climate modelling.

We assume a geometry as shown in Fig. 2.8 (left). The conservation of the energy of a thin spherical air layer (as a model for the atmosphere) is given approximately as:

$$4\pi R^2 h \rho c \frac{dT}{dt} = \pi R^2 (1-\alpha) S_0 - 4\pi R^2 \varepsilon \sigma T^4 , \qquad (2.1)$$

where the following quantities are used:

$R = 6371$ km	Radius of the Earth
$h = 8.3$ km	Vertical extent of the air layer
$\rho = 1.2$ kg m^{-3}	Density of air
$c = 1000$ J kg^{-1} K^{-1}	Specific heat of air
T	Globally averaged surface temperature
$\alpha = 0.3$	Planetary albedo (reflectivity)
$S_0 = 1367$ W m^{-2}	Solar constant
$\varepsilon = 0.6$	Planetary emissivity
$\sigma = 5.67 \cdot 10^{-8}$ W m^{-2} K^{-4}	Stefan–Boltzmann constant

Equation (2.1) states that the heat content of the global atmosphere (left) can be changed due to two processes (right). The equation is a statement on the conservation of energy. This model is therefore referred to as an energy balance model (EBM). The first term on the right-hand side is the energy flux of the (mainly short-wave) radiation coming from the Sun, reaching the Earth through a circular disk, reduced by the reflected part. The second term describes the (mainly long-wave) irradiance emitted from the complete Earth surface. This term is a *parameterisation* of a complex process not further described in this model. The parameterisation assumes that the long-wave radiation can be quantified by the classical grey body radiation with parameter ε (emissivity). We will illustrate the role of this parameter by an example.

Equation (2.1) is an ordinary, non-linear differential equation of 1st order for an unknown time-dependent variable $T(t)$, the globally averaged surface temperature. For simple cases, (2.1) can be solved analytically.

2.2 Point Model of the Radiation Balance

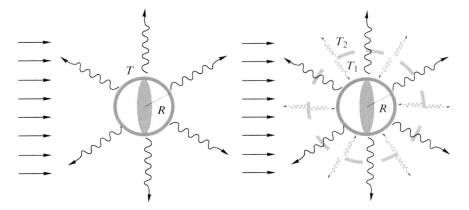

Fig. 2.8 Schematic depiction of simple global energy balance models (**left**), and of two radiating layers, resp. (**right**). The (mainly short-wave) radiation coming from the Sun is drawn with straight arrows; the (mainly long-wave) radiation from the Earth and from higher layers of the atmosphere is illustrated with wiggly arrows.

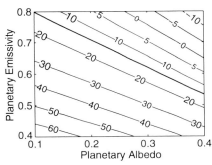

Fig. 2.9 Contour lines of equilibrium temperature according to (2.2) as a function of planetary albedo (α) and planetary emissivity (ε). The global mean surface temperature derived from measurements is equal to 14°C (*bold line*).

The equilibrium temperature can be found easily by setting the left-hand side equal to 0:

$$T = \left(\frac{(1-\alpha) S_0}{4 \varepsilon \sigma} \right)^{1/4}. \tag{2.2}$$

It is independent of the size of the Earth and the thermal characteristics of air. Figure 2.9 shows T in °C for different values of α and ε. The *bold line* highlights 14°C, approximately the mean surface temperature of the Earth. It is obvious that various, but not any, combination of the model parameters α and ε can yield 'realistic' solutions. The process of choosing model parameters in such a way that model results agree with nature, is called *tuning*. When tuning was applied, agreement of the model with observations is not a measure for the quality of the model unless further independent information about the values of *tunable parameters* is used.

In this case, estimates for α and ε based on remote sensing data (ERBE, *Earth Radiation Balance Experiment*) could be used to determine the components of the radiation balance. Results based on remote sensing yield a planetary albedo of

$\alpha = 0.3$. In order to obtain a mean temperature of 14°C in this EBM, the planetary emissivity has to be set to $\varepsilon = 0.6206$. This is a value significantly lower than the emissivity of natural surface areas which is around $\varepsilon \approx 0.8 \ldots 0.99$. Hence, this model parameter is unrealistic for an average Earth surface and does not give any information about the processes leading to this radiative equilibrium.

Assuming the Earth were a perfect black body, hence $\varepsilon = 1$, the temperature would be $-18.3°C$. Thanks to the natural greenhouse effect, mainly caused by water vapour and CO_2, we find a difference of approximately 32.3°C.

This will be illustrated with a second, slightly more complex EBM (Fig. 2.8, right). We assume, that irradiance occurs at the Earth surface at a temperature T_1, as well as from a higher level ("cirrus clouds", which are supposed not to affect the short wave radiation and hence the albedo) at temperature T_2. The high-altitude cloud cover is not complete, but extends over a fraction c of the total area. The stationary energy balance for both levels is given by:

$$\pi R^2 (1 - \alpha) S_0 + c 4\pi R^2 \sigma T_2^4 = 4\pi R^2 \varepsilon \sigma T_1^4 \tag{2.3a}$$

$$c 4\pi R^2 \varepsilon \sigma T_1^4 = 2c 4\pi R^2 \sigma T_2^4 , \tag{2.3b}$$

where we have assumed that the Earth's surface is a "grey" body with emissivity ε, the cloud cover is assumed to be a black body. The solution is now given as follows:

$$T_1 = \left(\frac{(1 - \alpha) S_0}{4\varepsilon\sigma \left(1 - \frac{c}{2}\right)} \right)^{1/4} \tag{2.4a}$$

$$T_2 = \left(\frac{(1 - \alpha) S_0}{4\sigma (2 - c)} \right)^{1/4} . \tag{2.4b}$$

Now, we have a slightly more detailed description of the "Earth's climate" (two temperatures). This comes at the expense of having more parameters (α, ε, c) for which reasonable values have to be chosen.

Figure 2.10 shows that in this model more realistic values of the surface emissivity can be applied. From Fig. 2.6 we derive a global-mean cloud cover of around 0.6. Tuning the model we choose $\varepsilon \approx 0.886$ and obtain an equilibrium temperature

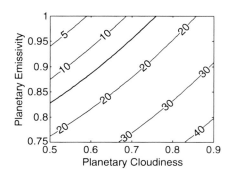

Fig. 2.10 Contour lines of equilibrium temperature according to (2.4a). The global mean surface temperature derived from measurements is equal to 14°C (*bold line*).

of 14°C. This yields $T_2 = -38.8°C$ the temperature that is approximately measured at a height of 8.2 km. An important information emerging from this model is that the Earth emits infrared radiation not only from the surface, but also from higher levels, as was already evident from Fig. 1.2. The natural greenhouse effect is caused by the fact that a higher irradiance occurs at a lower temperature and that these levels also radiate downwards (downward radiation). Hence, the surface is heated by a combination of direct short-wave solar radiation and long-wave back radiation. Figure 2.10 is only valid for high altitude clouds which do not affect α significantly. In general, clouds affect α and ε, and the net effect on a global scale is a cooling one (see Sect. 2.4.3).

In reality, the atmosphere has to be regarded as a continuum because radiative fluxes occur at all levels. These considerations lead to *radiative-convective models*, which are important components of AGCMs (category 1/0 in Table 2.1).

2.3 Numerical Solution of an Ordinary Differential Equation of First Order

We consider again the climate model given by (2.1) but now we examine its time-dependence. For this we will use a numerical algorithm.

Before we derive it, we look at the temporal behaviour of the energy balance model (EBM) near the equilibrium and write the temperature $T(t)$ as follows:

$$T(t) = \overline{T} + \tilde{T}(t) ,$$

where $\overline{T}$ is the constant equilibrium temperature given in (2.2) and $\tilde{T}$ is a small time-dependent temperature perturbation ($|\tilde{T}| \ll \overline{T}$). Hence, (2.1) can be written as

$$h\rho c \frac{d\tilde{T}}{dt} = \frac{1-\alpha}{4} S_0 - \varepsilon \sigma \left(\overline{T} + \tilde{T}\right)^4 . \tag{2.5}$$

Now we write $\left(\overline{T} + \tilde{T}\right)^4 = \overline{T}^4 \left(1 + \tilde{T}/\overline{T}\right)^4$ and use the Taylor series expansion

$$(1 + x)^n = 1 + nx + \frac{n(n-1)}{2} x^2 + \ldots$$

with $x = \tilde{T}/\overline{T}$ and $n = 4$. Neglecting the higher-order terms in this expansion with regard to $|\tilde{T}| \ll \overline{T}$ we obtain from (2.5), using (2.2)

$$\frac{d\tilde{T}}{dt} = -\left(\frac{4\varepsilon\sigma \overline{T}^3}{h\rho c}\right) \tilde{T} . \tag{2.6}$$

This is a linear, homogenous differential equation of 1st order for the temperature perturbation $\tilde{T}$, of which the solution is known:

$$\tilde{T}(t) = a\,e^{-t/\tau}, \qquad \tau = \frac{h\rho c}{4\varepsilon\sigma \overline{T}^3}, \tag{2.7}$$

where a is constant depending on the initial conditions ($a = \tilde{T}(0)$). Solution (2.7) states that a temperature disturbance in the EBM approximately decays on a characteristic time scale of $\tau \approx 35$ days and the radiation equilibrium is attained at temperature $T(t) = \overline{T}$. Hence, the temporal behaviour is determined by the thermal properties of the atmosphere and responds rather rapidly. Above considerations also show that $\overline{T}$ is a stable state, because the perturbation $\tilde{T}(t)$ approaches 0 for $t \to \infty$, as evident from (2.7).

In the following we will discuss the procedure to solve (2.1) numerically. First, the question arises of how to compute the derivatives in this equation. We assume that it is sufficient to know them only at certain points in time chosen a priori. Therefore, the problem can be discretized in time. The times are chosen according to the rule

$$t = n\,\Delta t, \qquad n = 0, 1, 2, \ldots \tag{2.8}$$

Δt is the *time step*. Equation (2.8) can also be interpreted as grid points on the time axis. It has to be noted that the time step has to be significantly shorter than the characteristic time scales of the processes described by the model. In the present case $\Delta t \ll 35$ days would be selected.

Let us assume we know the solution at time t. Therefore, the function $T(t)$ can be expanded in a Taylor series:

$$T(t + \Delta t) = T(t) + \left.\frac{dT}{dt}\right|_t \Delta t + \frac{1}{2!}\left.\frac{d^2T}{dt^2}\right|_t \Delta t^2 + \frac{1}{3!}\left.\frac{d^3T}{dt^3}\right|_t \Delta t^3 + \ldots. \tag{2.9}$$

We can solve (2.9) for the first derivative evaluated at time t:

$$\left.\frac{dT}{dt}\right|_t = \frac{T(t+\Delta t) - T(t)}{\Delta t} \underbrace{- \frac{1}{2!}\left.\frac{d^2T}{dt^2}\right|_t \Delta t - \frac{1}{3!}\left.\frac{d^3T}{dt^3}\right|_t \Delta t^2 - \ldots}_{\text{terms of order } \Delta t \text{ and higher}}. \tag{2.10}$$

By neglecting the terms of order Δt and higher we obtain the so-called *Euler scheme*, a finite difference scheme of 1st order. This means that the corrections of this scheme scale with Δt. Whether the scheme is correct can be directly determined by considering the limit $\Delta t \to 0$. It is the simplest but at the same time the most inaccurate way of calculating first derivatives.

Adding to (2.10) the corresponding equation with Δt replaced by $-\Delta t$, a new equation results which yields an alternative scheme for the first derivative:

$$\left.\frac{dT}{dt}\right|_t = \frac{T(t+\Delta t) - T(t - \Delta t)}{2\,\Delta t} \underbrace{- \frac{1}{3!}\left.\frac{d^3T}{dt^3}\right|_t \Delta t^2 - \frac{1}{5!}\left.\frac{d^5T}{dt^5}\right|_t \Delta t^4 - \ldots}_{\text{terms of order } \Delta t^2 \text{ and higher}}.$$

$$\tag{2.11}$$

2.3 Numerical Solution of an Ordinary Differential Equation of First Order

Table 2.2 Overview of the simplest schemes for the calculation of 1st and 2nd derivatives of the function f.

Continuous	Finite differences	Error	Name
$f'(x)$	$\dfrac{f(x+\Delta x) - f(x)}{\Delta x}$	$O(\Delta x)$	Euler forward
$f'(x)$	$\dfrac{f(x) - f(x - \Delta x)}{\Delta x}$	$O(\Delta x)$	Euler backward
$f'(x)$	$\dfrac{f(x+\Delta x) - f(x-\Delta x)}{2\Delta x}$	$O(\Delta x^2)$	centered difference
$f''(x)$	$\dfrac{f(x+\Delta x) - 2f(x) + f(x-\Delta x)}{\Delta x^2}$	$O(\Delta x^2)$	centered difference

This is the scheme of *centered differences*. The name refers to the position on the time grid, where derivatives at one point are calculated by taking differences of values from two neighbouring points. The corrections of this scheme scale with Δt^2 and for small Δt, they converge to 0 faster than in (2.10). These simple schemes are summarized in Table 2.2.

The formulations assume an equidistant discretization; adjustments are necessary if the grid's resolution is spatially dependent (e.g., on a spherical spatial grid).

We now solve (2.6) numerically by using the Euler forward scheme:

continuous: $\qquad \dfrac{dT}{dt} = -AT, \qquad T(t)$

discrete: $\qquad \dfrac{T_{n+1} - T_n}{\Delta t} = -AT_n, \qquad T_n \equiv T(n\Delta t)$

and obtain:

$$T_{n+1} = T_n - AT_n \Delta t = (1 - A\Delta t) T_n = \ldots = (1 - A\Delta t)^{n+1} T_0. \quad (2.12)$$

Is the numerical solution (2.12) consistent with the analytical solution (2.7)? We would like to show that for the limit of $\Delta t \to 0$, the numerical solution converges towards the analytical one. Therefore, we apply a transformation of variables $s = -1/(A\Delta t)$ and take the limit $s \to \infty$:

$$T(t) = T(n\Delta t) = T_n = T_0 (1 - A\Delta t)^n = T_0 \left(1 + \dfrac{1}{s}\right)^{-sAt}$$

$$= T_0 \left(\left(1 + \dfrac{1}{s}\right)^s\right)^{-At}. \quad (2.13)$$

The following is valid,

$$\lim_{s \to \infty} \left[T_0 \left(\left(1 + \frac{1}{s}\right)^s \right)^{-At} \right] = T_0 \, e^{-At},$$

in agreement with (2.7).

Hence, it has been shown that the numerical solution converges towards the analytical solution for arbitrarily small Δt. But are there some cases where the scheme would fail?

From (2.12) it can be derived that for $\Delta t = 1/A$ the scheme yields $T_n = 0$, whereas for $\Delta t = 2/A$ it yields $T_n = (-1)^n \, T_0$; both results do not make sense. This is a distinctive feature of the equation to be solved. Central differences may also cause general problems in the case, for example, that periodic solutions with unluckily chosen time steps should be calculated.

The Euler scheme is the simplest, but also the most inaccurate one-step scheme. Generally, it solves

$$\frac{dy}{dx} = f(x, y(x)) \tag{2.14}$$

with an initial condition $y(x_0) = y_0$. For the EBM given by (2.1) the following correspondences hold: $y = T$, $x = t$ and $f(x, y) = (1 - \alpha) \, S_0 / (4 \, h \, \rho \, c) - \varepsilon \sigma \, y^4 / (h \, \rho \, c)$. The Euler scheme evaluates derivatives only at the points x and $x + \Delta x$ which corresponds to the linearisation which was used in (2.10).

The evaluation of $f(x, y)$ at further locations in the interval $[x, x + \Delta x]$ and by a suitable linear combination, the error can be reduced from $O(\Delta x)$ to $O(\Delta x^k)$. This leads to schemes of the type *Runge–Kutta of order k*. For $k = 4$ we obtain the classical Runge–Kutta scheme for which the rule is as follows:

$$\begin{aligned}
y_{n+1} &= y_n + \Delta x \, F(x_n, y_n) \\
F(x_n, y_n) &= \tfrac{1}{6} \left(K_1 + 2 \, K_2 + 2 \, K_3 + K_4 \right) \\
K_1 &= f(x_n, y_n) \\
K_2 &= f\left(x_n + \tfrac{1}{2} \Delta x, \, y_n + \tfrac{1}{2} \Delta x \, K_1\right) \\
K_3 &= f\left(x_n + \tfrac{1}{2} \Delta x, \, y_n + \tfrac{1}{2} \Delta x \, K_2\right) \\
K_4 &= f\left(x_n + \Delta x, \, y_n + \Delta x \, K_3\right).
\end{aligned} \tag{2.15}$$

Figure 2.11 compares the different schemes with the exact solution (2.7) of the linearized system (*red line*). The Euler scheme was applied with time steps of $\Delta t = 12, 24, 36, 50$ days. Schemes, for which the time step is larger than their characteristic time scale τ, see (2.7), do converge to the exact solution but show a completely wrong transient behaviour. By using smaller time steps, the exact solution can be approximated with increasing accuracy. Only time steps smaller than the characteristic time scale of 35 days approximately yield the transient behaviour

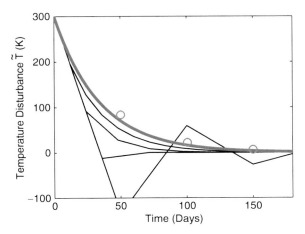

Fig. 2.11 Numerical solutions of (2.6) with the initial condition $\tilde{T}(0) = 300$ K computed with the Euler scheme and time steps of 12, 24, 36, 50 days. The exact solution of the linearized system is drawn in a *red line*, the results from the classical Runge–Kutta scheme ($\Delta t = 50$ days) are labelled with *green circles*.

of the exact solution when using the Euler forward scheme. For the Runge–Kutta scheme (circles) $\Delta t = 50$ days was chosen. The agreement with the exact solution is already significantly better than with the Euler scheme with $\Delta t = 12$ days, in spite of the large time step.

The use of the Runge–Kutta scheme requires that the function f in (2.14) can be evaluated at any point (x, y). In most of the climate models this important prerequisite is not fulfilled and the Runge–Kutta scheme can therefore not be applied for the time integration.

2.4 Climate Sensitivity and Feedbacks

An important quantity in climate dynamics is the *equilibrium climate sensitivity*, defined as the global mean temperature change resulting from a doubling of the atmospheric CO_2 concentration after the climate system has re-established a new equilibrium. This quantity, often referred to as $\Delta T_{2\times}$, is a fundamental characteristic of the climate system and at the same time a useful metric for climate models. It serves to compare models of different categories or of successive generations. Over the last three decades $\Delta T_{2\times}$ was estimated at 1.5–4.5°C, without any information about a possible distribution within this range, see IPCC (2001). In the latest IPCC report, IPCC (2007), more quantitative statements about the climate sensitivity could be made for the first time:

- Likely range (66%): 2–4.5°C.
- Very unlikely (< 10%): smaller than 1.5°C.
- Most likely value: around 3°C.

The *equilibrium climate sensitivity* is evaluated when the climate model has established a new equilibrium under an altered radiation balance. In expensive coupled climate models, it usually has to be determined by a temporal extrapolation.

The temperature increase with a doubling of the atmospheric CO_2 concentration is the result of complex processes and interactions in the atmosphere that affect the radiation balance. The contributions of the single processes as a response to the disturbance of the radiation balance (e.g., by an increase in greenhouse gas concentrations or a volcanic eruption) can be quantified by the strength of the feedback. Therefore, the term *feedback parameter*, given as λ (W m^{-2} K^{-1}), is introduced. It quantifies the change in the radiation balance per change of the global mean temperature. The estimation of λ for various processes is a central task of climate research.

The concept of *feedback parameters* can be illustrated using the linearised EBM. We write for the energy balance:

$$0 = A(T) + B(T) + W(T) + \Delta Q , \qquad (2.16)$$

where A is the short-wave radiation (which may be temperature-dependent via albedo), B is the long-wave back-radiation, W is an additional term of the radiation balance (e.g., effects of clouds, greenhouse gases, such as H_2O, CO_2, ..., aerosols, etc.) and ΔQ is a disturbance (often called *forcing*) of the balance which causes a change in temperatures and shall be determined.

We expand all functions of T around the equilibrium temperature $\overline{T}$ and obtain

$$\begin{aligned} 0 &= A(\overline{T}) + A' \cdot (T - \overline{T}) + B(\overline{T}) + B' \cdot (T - \overline{T}) + W(\overline{T}) + W' \cdot (T - \overline{T}) + \Delta Q \\ &= \underbrace{A(\overline{T}) + B(\overline{T}) + W(\overline{T})}_{=0} + (A' + B' + W') \cdot (T - \overline{T}) + \Delta Q , \end{aligned}$$

where A', B' and W' denote the first derivatives with respect to T of the functions $A(T)$, $B(T)$ and $W(T)$ at $T = \overline{T}$, respectively. We define the *feedback parameter* as

$$\lambda = \lambda_A + \lambda_B + \lambda_W = A' + B' + W' . \qquad (2.17)$$

Hence, the new temperature T is

$$T = \overline{T} - \frac{1}{\lambda} \Delta Q = \overline{T} + s \, \Delta Q , \qquad (2.18)$$

where

$$s = -\frac{1}{\lambda}$$

often is denoted the *sensitivity parameter* (K/(W m^{-2})). The smaller λ, the larger is the temperature change due to a perturbation ΔQ. The total feedback is the sum of the single feedbacks; the total sensitivity is equal to the inverse of the sum of the

2.4 Climate Sensitivity and Feedbacks

inverse sensitivities:

$$\lambda = \lambda_A + \lambda_B + \lambda_W, \qquad \frac{1}{s} = \frac{1}{s_A} + \frac{1}{s_B} + \frac{1}{s_W}. \qquad (2.19)$$

In the following this will be applied to the "two-layer"-EBM presented in (2.3). The radiation budget for the surface temperature is given by

$$0 = \frac{1-\alpha}{4} S_0 - \varepsilon \sigma T^4 + \underbrace{\frac{c}{2} \varepsilon \sigma T^4}_{W},$$

where W describes the effects of high clouds. Cirrus clouds cause a positive contribution to the radiation balance, hence a warming.

The derivatives of the individual radiation terms yield the individual *feedback parameters*:

$$\lambda_A = -\frac{S_0}{4}\frac{d\alpha}{dT}, \quad \lambda_B = -4\varepsilon\sigma T^3, \quad \lambda_W = \frac{c}{2}4\varepsilon\sigma T^3 + \frac{1}{2}\varepsilon\sigma T^4 \frac{dc}{dT}. \quad (2.20)$$

Assuming that the albedo is not temperature-dependent, and the effect of high clouds is irrelevant, and no additional forcing exists, we obtain:

$$\lambda = \lambda_B = -4 \cdot 0.6206 \cdot 5.67 \cdot 10^{-8} \cdot (287.15)^3 \text{ W m}^{-2}\text{ K}^{-1}$$
$$= -3.33 \text{ W m}^{-2}\text{ K}^{-1}. \qquad (2.21)$$

This is the feedback parameter of long-wave radiation without other feedbacks, in particular without the water vapour feedback. This is referred to as the *Planck feedback*, also denoted λ_P. The feedback is negative, implying that an increase in temperature leads to an increased long-wave irradiance and hence to a cooling. Latest estimates from various climate models yield $\lambda_P = -(3.21 \pm 0.04)$ W m^{-2} K^{-1} (Soden and Held 2006).

Especially the strong temperature dependence of the water vapour content in the atmosphere (via the Clausius–Clapeyron equation) – the most important greenhouse gas – as well as the temperature-dependent change in albedo and cloud cover, strongly affect the overall feedback. We would like to assess this with the ice-albedo feedback, the water vapour- and the cloud feedback.

2.4.1 Ice-Albedo Feedback

A globally and locally important feedback mechanism arises from the temporal and spatial change in the extent of the snow- and ice cover with changing temperatures. If the extent of the snow and ice cover is large – this is generally the case at low temperatures – more solar radiation is reflected. Snow and ice have a high

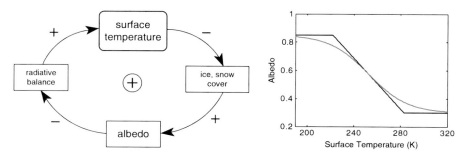

Fig. 2.12 Ice-albedo feedback (**left**) and two plausible parameterisations for an EBM (**right**). The signs next to the arrows denote the correlation between changes in the quantities in the boxes at the beginning and at the end of the arrow. The resulting correlation is given in the centre of the *feedback loop*. A self-enhancing process has a positive sign and can therefore lead to instabilities in the climate system. A negative sign corresponds to a damped process. For the parameterisation shown in the graph at the right, it is assumed that for low temperatures a complete snow-/ice cover exists and the albedo is $\alpha \approx 0.85$. For high temperatures a planetary albedo of 0.3 is assumed.

reflectivity, i.e., albedo ($\alpha \approx 0.85$). This implies a positive feedback, as one can see in Fig. 2.12 (left). Under a global warming scenario the extent of the snow and ice cover is expected to shrink; also, the seasonal snow and ice cover begins later and ends earlier. This leads to a shortening of the seasonal cover and hence to a positive contribution to the seasonal radiation balance.

This temperature-dependence of the albedo shall be parametrised in the EBM. This problem was studied by Sellers as early as 1969 (Sellers 1969), who based it on the parameterisation given in Fig. 2.12 (right). It is obvious that in a global model the evolution of the snow and ice cover cannot be simulated. For this reason, *plausible* assumptions are made, which may be based on the correlation of snow cover and regional temperatures. Sellers proposed:

$$\alpha = 0.3 - 0.009\,(T - 283\,\text{K})/\text{K}\,, \qquad 222\,\text{K} \leq T \leq 283\,\text{K}\,. \tag{2.22}$$

A mathematically differentiable function may be preferable (Fig. 2.12, right).
From (2.22) we derive

$$\lambda_A = -\frac{S_0}{4}\frac{d\alpha}{dT} = \frac{1367 \cdot 0.009}{4}\,\text{W}\,\text{m}^{-2}\,\text{K}^{-1} = 3.08\,\text{W}\,\text{m}^{-2}\,\text{K}^{-1}\,, \tag{2.23}$$

hence, again a positive feedback. Therefore, the total feedback is

$$\lambda = \lambda_B + \lambda_A = (-3.33 + 3.08)\,\text{W}\,\text{m}^{-2}\,\text{K}^{-1} = -0.25\,\text{W}\,\text{m}^{-2}\,\text{K}^{-1}\,. \tag{2.24}$$

Compared with (2.21) this results in a large reduction of the absolute value of the feedback parameter which causes a strong enhancement of the sensitivity. An approach resulting in (2.24) is still far from reality since not the whole planet but only polar regions may be exposed to such a process. Latest estimates with various climate models yield $\lambda_A = (0.26 \pm 0.08)$ W m^{-2} K^{-1} (Bony et al. 2006).

The fact that (2.24) must be too large is supported by estimates for the glacial-interglacial temperature difference. Nevertheless, this examination is instructive in the sense that it illustrates the effect of positive and negative feedbacks and their combination.

2.4.2 Water Vapour Feedback

The water vapour feedback is the most important feedback in the climate system because water vapour is the primary natural greenhouse gas. A warm atmosphere can hold more water vapour than a cold atmosphere. These additional water molecules in the warm atmosphere cause an enhancement of the natural greenhouse effect by increased absorption of long-wave radiation. Latest estimates from various climate models yield $\lambda_{WV} = (1.80 \pm 0.18)$ W m^{-2} K^{-1} (Bony et al. 2006).

With this we find

$$\lambda = \lambda_B + \lambda_{WV} = (-3.33 + 1.8) \text{ W m}^{-2}\text{ K}^{-1} = -1.53 \text{ W m}^{-2}\text{ K}^{-1}, \qquad (2.25)$$

hence, again significant reduction of the absolute value of λ which amounts to an increased sensitivity by a factor of 2 compared to (2.21). *The presence of water vapour in the atmosphere doubles the climate sensitivity.*

It is difficult to directly observe the water vapour feedback, but various independent approaches have resulted in a much better quantification of this feedback in the last few years. The agreement of the spatial structure of the water vapour distribution, as it was shown in Fig. 2.3, does not yet guarantee that climate models compute the climate sensitivity in a reasonable way.

However, based on observations of the change in temperature after the large volcanic eruption of Pinatubo in 1991, it has been shown that current climate models simulate the water vapour feedback reasonably well. A climate model with water vapour feedback is capable of simulating the global cooling of the mid-troposphere by 0.7°C following the eruption (Fig. 2.13). A model, in which the water vapour content was fixed, shows a significantly smaller cooling. Such a model therefore has a smaller sensitivity as expected from (2.25). Figure 2.13 also points to the fact that current climate models simulate this effect rather well.

2.4.3 Cloud Feedback

Modelling the cloud cover still belongs to one of the greatest challenges in climate modelling and in the assessment of future climate change. A fundamental aspect of the problem is apparent in Fig. 2.14. It illustrates, in a very simplified form, two

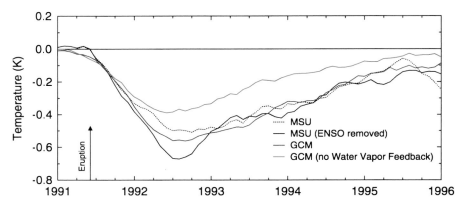

Fig. 2.13 Global mean temperature anomaly in the mid-troposphere after the eruption of Mount Pinatubo in 1991. A global cooling of 0.7°C was observed with remotely sensed radiation measurements (microwave sounding unit, MSU) after a warming effect of the 1992/1993 ENSO was subtracted. A climate model in which the water vapour feedback was turned off shows a smaller cooling inconsistent with the observations. Figure from Soden et al. (2002).

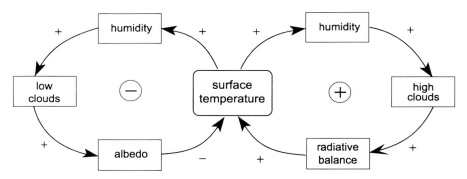

Fig. 2.14 Cloud feedback loops. The sign depends on the location and the quality of clouds. Low clouds affect short-wave radiation via albedo as opposed to high clouds affecting long-wave irradiance.

possible feedback mechanisms: They can be positive or negative because clouds affect both short-wave radiation (via albedo) and long-wave radiation.

A global estimate for the effect of clouds is given in Table 2.3. The averaged effect of the global cloud cover results in a *cooling* which suggests the albedo effect dominates. The estimates in Table 2.3 yields a value for the *forcing* with respect to the change in cloud cover, under the assumption of a mean cloud cover of 60%, of about

$$\frac{\Delta W}{\Delta \text{Clouds}} \approx -\frac{17\,\text{W}\,\text{m}^{-2}}{60\%} \approx -0.3\,\text{W}\,\text{m}^{-2}/\% \,. \tag{2.26}$$

An increase in cloud cover by about 13% would constitute a forcing of $\Delta W \approx -3.7\,\text{W}\,\text{m}^{-2}$. This negative forcing (cooling) would compensate the positive forcing expected from a doubling of the atmospheric CO_2 concentration (see (2.31) below).

2.4 Climate Sensitivity and Feedbacks

Table 2.3 Estimate for the change in radiation in W m^{-2} due to the global cloud cover (from Hartmann 1994).

	Mean	Without clouds	With clouds
Long-wave radiation	−234	−266	+31
Absorbed short-wave radiation	239	288	−48
Net radiation	**+5**	**+22**	**−17**
Albedo	30%	15%	+15%

To illustrate the concept, consider the two-layer EBM given by (2.3) as a model for a very simplified representation of the effect of clouds and assume – as a first step – that c does not depend on the temperature and $c \approx 0.6$ (Fig. 2.6). Hence, (2.21) becomes

$$\lambda = \lambda_B + \lambda_W = (-3.33 + 1.0) \text{ W m}^{-2}\text{ K}^{-1} = -2.33 \text{ W m}^{-2}\text{ K}^{-1} \quad (2.27)$$

which suggests a reduction of the absolute value of λ, corresponding to an increase in the sensitivity ($\approx 50\%$) compared to (2.21).

Of course, the two-layer EBM is not a realistic model to quantify the cloud feedback correctly. To this end, atmosphere models are necessary that resolve the formation of clouds in all their forms. Latest estimates from several climate models yield $\lambda_W = (0.69 \pm 0.38)$ W m^{-2} K^{-1} (Bony et al. 2006).

Within the last few years model consistency with regard to the cloud feedback has increased considerably. Multiple lines of evidence indicate that the total feedback is *positive*. A warming by 1°C leads to a total additional forcing (change in cloud cover, structure of the clouds, albedo and height of clouds) of about 0.7 W m^{-2}.

2.4.4 Lapse Rate Feedback

A warming trend also modifies the vertical structure of the atmosphere because the warming is not necessarily uniform over the air column. Rather, different rates of warming may occur at different altitudes. This leads to another feedback, called *lapse rate feedback*. The lapse rate (of the atmospheric air temperature) is defined as the rate of decrease of the atmospheric air temperature T with increase in altitude z,

$$\gamma = -\frac{\partial T}{\partial z}.$$

It amounts to about 10°C km^{-1} in a dry atmosphere and to about 6°C km^{-1} in a humid atmosphere.

The lapse rate strongly depends on the location. In the tropics, a warming leads to an increased convective activity: water vapour rises and condensates at high altitudes. This transport of latent heat results in a stronger warming in the high layers of the atmosphere which is supported by the additional greenhouse effect due to the increased concentration of water vapour there. In consequence, the lapse rate

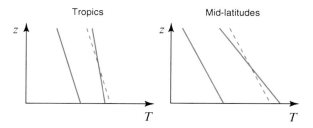

Fig. 2.15 Schematic illustration of changes in the vertical temperature structure in the tropics and in the mid-latitudes. Due to increased convection, the lapse rate decreases in the tropics. In the mid-latitudes the horizontal flow limits the warming to the surface and the lower atmosphere which causes the lapse rate to increase. Compared to the mean warming, a reduced warming of the surface occurs in the tropics (therefore a negative feedback), while it is enhanced in the mid-latitudes (positive feedback).

decreases (Fig. 2.15, left panel). In the mid-latitudes, where horizontal circulation associated with high- and low-pressure systems dominates, and hence, the vertical movement is less pronounced compared to the tropics, the warming is limited to layers close to the surface. This leads to an enhanced lapse rate (Fig. 2.15, right panel). With regard to the surface temperatures, the lapse rate feedback is therefore negative in the tropics, whereas it is positive in the mid-latitudes. For the global average the tropics dominate due to their larger spatial extent. The resulting feedback is therefore negative but with rather large uncertainties. Latest estimates from several climate models yield $\lambda_{LR} = (-0.84 \pm 0.26)$ W m^{-2} K^{-1} (Bony et al. 2006).

2.4.5 Summary and Conclusion Regarding Feedbacks

Figure 2.16 summarizes the various feedbacks discussed above. Different model studies and the inclusion of remote sensing data, as well as direct measurements permit a quantification of the single feedbacks. The strongest positive feedback is the water vapour feedback, which – in spite of the negative lapse rate feedback – remains positive in total. The finding that the cloud feedback is positive in total is new but still is associated with the largest uncertainties.

The best estimate for the *Planck-feedback* is $\lambda_P = -3.2$ W m^{-2} K^{-1}. Therefore, the total feedback becomes:

$$\lambda = \lambda_P + \lambda_{all} = (-3.2 + 1.9) \text{ W m}^{-2} \text{ K}^{-1} = -1.3 \text{ W m}^{-2} \text{ K}^{-1}. \qquad (2.28)$$

With this, the equilibrium climate sensitivity $\Delta T_{2\times}$ can be estimated. We write

$$\Delta T_{2\times} = \frac{-1}{\lambda} \Delta Q_{2\times}, \qquad (2.29)$$

2.4 Climate Sensitivity and Feedbacks

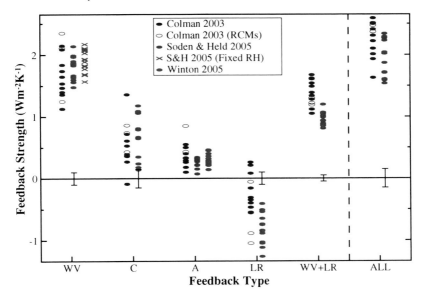

Fig. 2.16 Overview of the most important feedbacks in the atmosphere: WV (water vapour feedback), C (cloud feedback), A (albedo feedback), LR (lapse rate feedback), ALL (total feedback). Thanks to the latest remote sensing and (radiosonde) measurements, the sum of WV+LR can be estimated more precisely than the single components. The total feedback is clearly positive. Vertical bars indicate the uncertainty in the calculations of Soden and Held (2006). Figure from Bony et al. (2006).

where $\Delta Q_{2\times}$ denotes the *forcing* caused by a doubling of the atmospheric CO_2 concentration.

The radiative forcing associated with changes in the atmospheric concentration of CO_2 is given by Myhre et al. (1998):

$$\Delta Q(CO_2) = 5.35 \, \text{W m}^{-2} \ln \frac{[CO_2]}{280 \, \text{ppm}}, \tag{2.30}$$

hence

$$\Delta Q_{2\times} = 5.35 \, \text{W m}^{-2} \ln \frac{560 \, \text{ppm}}{280 \, \text{ppm}} = 3.7 \, \text{W m}^{-2}. \tag{2.31}$$

From (2.29) it follows, that

$$\Delta T_{2\times} = \frac{-1}{\lambda} \Delta Q_{2\times} = 2.85 \, \text{K}. \tag{2.32}$$

This is close to the best estimate of the equilibrium climate sensitivity $\Delta T_{2\times}$, as given by IPCC (2007) and reviewed by Knutti and Hegerl (2008).

The combined effect of different feedbacks can be illustrated by a latitudinal and altitudinal cross-section of the warming of the atmosphere with an increase in

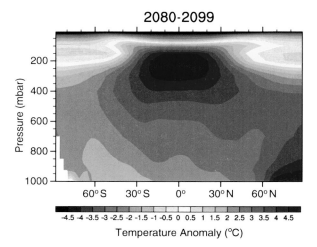

Fig. 2.17 Mean of 21 AOGCM models which were used for the projection of climate change for the Fourth Assessment Report of the Intergovernmental Panel on Climate Change, IPCC (2007). The zonal average temperature change in the atmosphere over the last 20 years of the twenty-first century for the emission scenario A1B is shown. The troposphere shows a warming, amplified between 7 and 12 km in the tropics, as well as on the surface north of 60°N. The stratosphere shows a cooling. Figure from IPCC (2007), Chap. 10 (Fig. 10.7, p. 765).

CO_2 concentrations. Figure 2.17 shows the zonal mean temperature change in the years 2080–2099 in a multi-model ensemble for the emission scenario A1B. The warming is stronger at high latitudes of the northern hemisphere towards the surface. This increase is caused by the *ice-albedo feedback*, which is mainly effective in the northern hemisphere, where the seasonal snow cover undergoes fast changes.

A clear enhancement of the warming also occurs in latitudes between 30°S and 30°N at an altitude between 7 and 12 km. This is due to the lapse rate feedback which leads to an increase in the water vapour content and hence to the concentration of the most important greenhouse gas in these altitudes. This is due to the temperature-dependence of the moist adiabatic curve: At higher temperatures, the lapse rate decreases. Therefore, the temperature rises faster at high altitudes (Fig. 2.15).

An important *fingerprint* of global warming is expected to take place in the stratosphere, where a cooling will occur at all latitudes. This cooling is actually observed (IPCC (2007), Chap. 3, Fig. 3.17). It is due to the rise of the irradiance altitude for long-wave radiation with an increase in CO_2 concentrations. At these higher altitudes, the temperatures are lower (in equilibrium at $T \approx 255$ K, hence at 5.1 km). This causes a disequilibrium, which the warming of the whole atmosphere compensates for. This warming leads to a rise of the irradiance altitude (level of equivalent black body radiation). Hence, a bigger part of the atmosphere now lies underneath the irradiance altitude, meaning that the optical path up to the radiation altitude has increased. Underneath this altitude a larger part of the long-wave irradiance is

2.4 Climate Sensitivity and Feedbacks

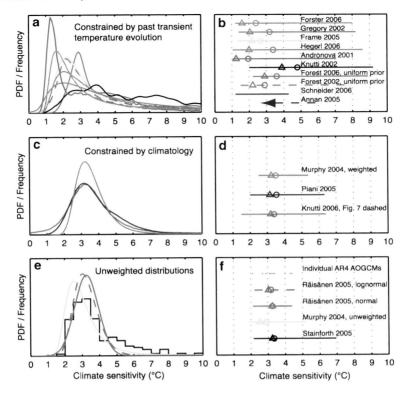

Fig. 2.18 (a) Probability distribution from studies determining the climate sensitivity based on past climate changes. (b) As in panel (a), but 5%–95% ranges, median (*circles*) and most probable value (*triangle*). (c)/(d) As (a)/(b) but based on the climatology. (e)/(f) Unweighted or adjusted distributions of different models or of a model with different parameters. Figure from IPCC (2007), Chap. 10 (Box 10.2, Fig. 1, p. 798).

absorbed and the stratosphere experiences a corresponding deficit, which leads to a cooling.

Thanks to a significantly improved knowledge of the individual feedback mechanisms in the atmosphere, the equilibrium climate sensitivity $\Delta T_{2\times}$ is now better quantified. Climate models of different categories of the hierarchy (Table 2.1) are used to simulate the temperature change over the last 150–1,000 years. The agreement of the model simulations with observations and paleo-reconstructions is computed which provides constraints for the range of various tuning parameters in the models, or eliminates certain simulations.

In summary, this yields an estimate of the probability distribution of the climate sensitivity, as it is shown in Fig. 2.18. On this basis IPCC (2007) concluded that the equilibrium climate sensitivity has a most probable value of around 3°C, as mentioned at the beginning of Sect. 2.4.

Chapter 3
Describing Transports of Energy and Matter

In nature the transport of energy and matter in fluids is determined by diffusion and advection. These processes induce fluxes of energy and matter, of which the mathematical description is derived by continuum mechanics. Diffusion is a random process taking place at all times and leading to a net transport only under certain conditions. Advection is caused by an ambient flow which transports energy and matter.

All processes in the climate system are fundamentally influenced by the advective and diffusive transport of mass, energy, momentum. For example, the temperatures at a particular latitude are determined by the balance of heat at that location which consists of the local radiation fluxes and the horizontal transport of heat in the atmosphere, including the transport of moisture. Another example concerns the transport of salt in the ocean through advective and diffusive processes. These change the density and are thus exerting a strong influence on the large-scale circulation in the ocean. Hence, the mathematical descriptions of these transport processes in models is fundamental to climate science.

3.1 Diffusion

Diffusive processes are caused by the thermal motion of molecules (Brownian motion) and can be described only in a statistical way. We consider first the one-dimensional case and divide the x-axis into cells of width Δx and cross-section area A in which molecules reside (Fig. 3.1).

Due to a positive thermodynamic temperature $T > 0$ the molecules are in thermal motion. The particle density (particles per volume) at coordinate x is denoted by $n(x)$. We describe the random motion by a probability p that a particle jumps from one cell to the neighboring cell. We further assume, that diffusion is an isotropic process (this is not always the case in nature). Therefore, the probability p is uniform and independent of the direction of the particle movement.

We determine the particle flux density (particles per unit area and unit time) at the cell boundary $i/i+1$ for a time interval Δt. From cell i, a number of $pn(x_i)A\Delta x$ particles jump to the right, while from cell $i+1$ a number $pn(x_i + \Delta x)A\Delta x$ jump

Fig. 3.1 Model of one-dimensional diffusion. The particle density in cell i is given by $n(x_i)$.

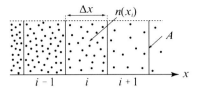

to the left. Hence, the net diffusive particle flux density (number of particles per area A and time Δt) at the cell boundary $i/i+1$ is given by

$$F = \frac{pn(x_i)A\Delta x - pn(x_i+\Delta x)A\Delta x}{A\Delta t} = -\frac{p\Delta x^2}{\Delta t}\frac{n(x_i+\Delta x)-n(x_i)}{\Delta x}.$$

In the limit of $\Delta x \to 0$ and $\Delta t \to 0$, provided $\Delta x^2/\Delta t =$ constant, we get *Fick's first law* of one-dimensional diffusion

$$F = -\left(\frac{p\Delta x^2}{\Delta t}\right)\frac{\partial n}{\partial x} = -D\frac{\partial n}{\partial x}. \tag{3.1}$$

The quantity D is the *diffusion constant*, also referred to as *diffusion coefficient* or *diffusivity*, with the unit m² s⁻¹; it depends on the physical properties of both the diffusing particles and the medium containing these particles (the medium can be vacuum, a gas, a liquid or a solid). This derivation shows that the diffusion constant parametrises processes that evolve on a molecular scale.

From (3.1) it follows that net diffusive fluxes only occur when concentration gradients, in the case of (3.1) particle density gradients, are present. Due to the random motion, gross-fluxes of particles always exist.

The generalization of (3.1) to a three-dimensional isotropic space and a concentration $C = C(x, y, z)$ of an arbitrary physical quantity yields Fick's first law of three-dimensional diffusion

$$\vec{F} = -D\,\vec{\nabla}C. \tag{3.2}$$

$\vec{\nabla}$ is the gradient operator and D an isotropic diffusion constant (scalar). The gradient operator is given by

$$\vec{\nabla} = \left(\frac{\partial}{\partial x}, \frac{\partial}{\partial y}, \frac{\partial}{\partial z}\right)$$

and converts any scalar $\Phi(x, y, z)$, for example the concentration $C = C(x, y, z)$ from (3.2), into the gradient of $\Phi(x, y, z)$, the vector

$$\vec{\nabla}\Phi(x,y,z) = \left(\frac{\partial}{\partial x}, \frac{\partial}{\partial y}, \frac{\partial}{\partial z}\right)\Phi(x,y,z)$$

$$= \left(\frac{\partial \Phi(x,y,z)}{\partial x}, \frac{\partial \Phi(x,y,z)}{\partial y}, \frac{\partial \Phi(x,y,z)}{\partial z}\right),$$

3.2 Advection

Table 3.1 Examples of diffusive flux densities, ρ denotes a mass density and ρ_s the particle or the mass density of salt.

transported quantity	formulation
mass	$\vec{F} = -D\vec{\nabla}\rho$
heat	$\vec{F} = -D\rho c \vec{\nabla} T = -\lambda \vec{\nabla} T$
salt	$\vec{F} = -D\vec{\nabla}\rho_s$
y-momentum	$\vec{F} = -D\vec{\nabla}(\rho u_y)$

which points in the direction of the highest increase of $\Phi(x, y, z)$. The negative sign in (3.2) ensures, that the diffusive flux density $\vec{F}$ is in the opposite direction of the gradient, namely in the direction of the highest decrease of C. The diffusive flux densities in Table 3.1 serve as examples.

3.2 Advection

For the derivation of a formulation of advective flux densities of physical quantities in the climate system, we first consider the one-dimensional case which is illustrated in Fig. 3.2. We assume a flow $u(x, t)$ which transports the quantity to be considered. The fluid (gas, air, water) moves across a fixed control area A. The transported physical quantity (particles, mass, energy, momentum, tracer) is given as a concentration $C(x, t)$, hence, the quantity is referred to a volume. In a short time interval Δt a volume $A\Delta x$ of length $\Delta x = u\Delta t$ passes through a cross section of area A and transports the quantity $A\Delta x C$ through here. The advective flux density is given by

$$F = \frac{A \Delta x C}{A \Delta t} = \frac{\Delta x}{\Delta t} C = uC.$$

In three dimensions, the advective flux density of a scalar quantity C in a three-dimensional flow $\vec{u}(\vec{x}, t)$ is

$$\vec{F} = \vec{u}C. \tag{3.3}$$

The advective flux density is a vector aligned parallel to the flow. The advective flux densities in Table 3.2 serve as examples.

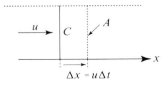

Fig. 3.2 Flow along the x-axis.

Table 3.2 Examples of advective flux densities, ρ denotes a mass density and ρ_s the particle or the mass density of salt.

transported quantity	formulation
mass	$\vec{F} = \vec{u}\rho$
heat	$\vec{F} = \vec{u}\rho c T$
salt	$\vec{F} = \vec{u}\rho_s$
y-momentum	$\vec{F} = \vec{u}\rho u_y$

3.3 Advection-Diffusion Equation and Continuity Equation

In the following discussion we will describe the connection between fluxes of physical quantities and time rates of changes of these quantities. It is established by formulating balance statements for those physical quantities which satisfy conservation laws. An example was presented in Sect. 2.2, where we have discussed a point model of the radiation balance.

We will set up a conservation equation for a physical quantity (for example number of particles, energy, mass, ...) with density C (i.e., particle density, energy density, mass density, ...) and start with one single dimension x (Fig. 3.3).

We consider a small fixed control volume $\Delta V = A \Delta x$. The (mean) density C inside the control volume changes in time due to fluxes into the control volume, fluxes out of the control volume and sources and sinks operating inside the control volume. Thus, we have

$$\frac{\partial}{\partial t}(C \Delta V) = F(x)A - F(x + \Delta x)A + P \Delta V, \tag{3.4}$$

where F is the flux density of quantity C and P is the *net source density* (sources minus sinks per unit volume) of this quantity. Inserting (3.2) and (3.3) into (3.4) and division by ΔV yields

$$\frac{\partial C}{\partial t} = -\frac{u(x+\Delta x)C(x+\Delta x) - u(x)C(x)}{\Delta x} + \frac{D\frac{\partial C}{\partial x}\big|_{x+\Delta x} - D\frac{\partial C}{\partial x}\big|_x}{\Delta x} + P$$

and taking the limit $\Delta x \to 0$, we obtain

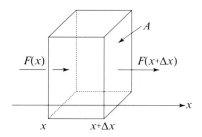

Fig. 3.3 Spatially dependent flux in one dimension.

$$\frac{\partial C}{\partial t} = -\frac{\partial (uC)}{\partial x} + \frac{\partial}{\partial x}\left(D\frac{\partial C}{\partial x}\right) + P. \tag{3.5}$$

Generalizing to three dimensions leads to the *advection-diffusion equation*:

$$\frac{\partial C}{\partial t} = -\vec{\nabla}\cdot(\vec{u}C) + \vec{\nabla}\cdot\left(D\,\vec{\nabla}C\right) + P, \tag{3.6}$$

where $\vec{\nabla}\cdot$ is the divergence operator. It acts on vectors and yields the "scalar product" of $\vec{\nabla}$ and the vector:

$$\vec{\nabla}\cdot\vec{u} = \left(\frac{\partial}{\partial x}, \frac{\partial}{\partial y}, \frac{\partial}{\partial z}\right)\cdot\begin{pmatrix}u_x\\u_y\\u_z\end{pmatrix} = \frac{\partial u_x}{\partial x} + \frac{\partial u_y}{\partial y} + \frac{\partial u_z}{\partial z}.$$

When C is the mass density and diffusion as well as sources or sinks vanish, then a special case arises from (3.6):

$$\frac{\partial \rho}{\partial t} = -\vec{\nabla}\cdot(\vec{u}\rho). \tag{3.7}$$

This is the *mass balance equation*, namely the general form of the continuity equation. Physically, (3.7) describes the conservation of mass: the total mass is conserved, mass is neither produced nor destroyed ($P = 0$), local mass density changes are always due to divergences of the mass flux (apart from molecular fluctuations due to diffusion). Equations (3.6) and (3.7) are balances representing the basis for the mathematical description of processes in the climate system. Their solution is the task of climate modelling.

For an incompressible fluid (e.g., ocean water in a thin interior layer) the density is constant and (3.7) simplifies to the continuity equation for incompressible fluids:

$$\vec{\nabla}\cdot\vec{u} = 0. \tag{3.8}$$

3.4 Describing Small- and Large-Scale Motions

The motions of the air in the atmosphere and of the water in the oceans can be very complex in detail. They are described using methods of geophysical fluid dynamics. The way this is achieved strongly depends on the spatial scale and the time scale. A useful concept is the statistical description of fluid flow.

Figure 3.4a shows an illustrative time series of wind velocity measurements, which could have been taken at a fixed position in the free atmosphere during a time of, for example, a few minutes, or a few days, or a few weeks. It illustrates the well-known consequences of the complexity just mentioned, namely a typically

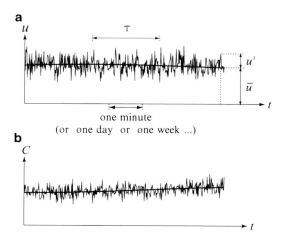

Fig. 3.4 Illustration of eddy fluctuations of (**a**) air velocity u and (**b**) another physical quantity C, for example particle density or humidity. *Bold lines* designate mean values ($\bar{u}$ and $\bar{C}$), τ is the averaging time.

slowly varying mean air velocity (denoted by the thick line in this figure) and a mostly rapidly varying deviation from the mean of the instantaneous air velocity. The cause for such a local time dependence of the air velocity are specific movements of numerous *eddies* of various sizes. These eddies are parts of the large air stream moving with the mean air velocity, mostly parts of larger eddies themselves, and move through the air surrounding them, after being released by irregular disturbances. In so doing they cause collectively so called *eddy fluctuations* of the air velocity at a point, i.e. local time varying deviations from the mean of the air velocity, and furthermore – as they transport advectively measurable air properties (e.g., water, CO_2, ...) – eddy fluctuations of the physical quantities C of these properties (Fig. 3.4b).

Climate research is mainly interested in processes on large spatial scales (global or continental) and long time scales (several days or longer). So the question arises whether the small and fast movements of the eddies within the large air stream have any relevance for the long-term trend of the physical quantity C satisfying the advection-diffusion equation (3.5). In the following we show that they have an influence and cannot be neglected in general. Consider the one-dimensional advection-diffusion equation (3.5) and separate the air velocity u, the physical quantity C and the source P in a temporal mean taken over successive time intervals $\tau = t_2 - t_1$ (which should be significantly shorter than the characteristic time scale of the processes to be considered) and an instantaneous deviation from this temporal mean, respectively,

$$u = \bar{u} + u', \qquad \bar{u} = \frac{1}{\tau} \int_{t_1}^{t_2} u(t)\,dt \qquad (\tau = t_2 - t_1)$$

$$C = \bar{C} + C', \qquad \bar{C} = \frac{1}{\tau} \int_{t_1}^{t_2} C(t)\,dt$$

3.4 Describing Small- and Large-Scale Motions

$$P = \overline{P} + P', \qquad \overline{P} = \frac{1}{\tau}\int_{t_1}^{t_2} P(t)\,dt,$$

where u', C' and P' denote instantaneous deviations from the time means $\overline{u}$, $\overline{C}$ and $\overline{P}$, just eddy fluctuations. The time means of the eddy fluctuations vanish, for example the time mean of the eddy fluctuation u':

$$\overline{u'} = \frac{1}{\tau}\int_{t_1}^{t_2} u'(t)\,dt = \frac{1}{\tau}\int_{t_1}^{t_2} (u(t) - \overline{u})\,dt = \overline{u} - \overline{u} = 0. \tag{3.9}$$

With this (3.5) becomes

$$\frac{\partial\left(\overline{C} + C'\right)}{\partial t} = -\frac{\partial\left((\overline{u} + u')\left(\overline{C} + C'\right)\right)}{\partial x} + \frac{\partial}{\partial x}\left(D\frac{\partial\left(\overline{C} + C'\right)}{\partial x}\right) + \overline{P} + P'.$$

Multiplying out the first term on the right side of the equation and using the sum rule of differentiation we obtain

$$\frac{\partial\overline{C}}{\partial t} + \frac{\partial C'}{\partial t} = -\frac{\partial\left(\overline{u}\,\overline{C}\right)}{\partial x} - \frac{\partial\left(u'\overline{C}\right)}{\partial x} - \frac{\partial\left(\overline{u}C'\right)}{\partial x} - \frac{\partial\left(u'C'\right)}{\partial x}$$

$$+ \frac{\partial}{\partial x}\left(D\frac{\partial\overline{C}}{\partial x}\right) + \frac{\partial}{\partial x}\left(D\frac{\partial C'}{\partial x}\right) + \overline{P} + P'.$$

This equation describes the processes at any moment exactly. But now we take the average with respect to time over the time interval (averaging time) τ, taking into account relation (3.9) and its consequences, namely

$$\overline{\frac{\partial\overline{C}}{\partial t}} = \frac{\partial\overline{\overline{C}}}{\partial t} = \frac{\partial\overline{C}}{\partial t}, \quad \overline{\frac{\partial\left(\overline{u}\,\overline{C}\right)}{\partial x}} = \frac{\partial\left(\overline{u}\,\overline{C}\right)}{\partial x}, \quad \overline{\frac{\partial}{\partial x}\left(D\frac{\partial\overline{C}}{\partial x}\right)} = \frac{\partial}{\partial x}\left(D\frac{\partial\overline{C}}{\partial x}\right),$$

$$\overline{\overline{P}} = \overline{P},$$

$$\overline{\frac{\partial C'}{\partial t}} = \frac{\partial\overline{C'}}{\partial t} = 0, \quad \overline{\frac{\partial\left(u'\overline{C}\right)}{\partial x}} = \overline{\frac{\partial\left(\overline{u}C'\right)}{\partial x}} = 0, \quad \overline{\frac{\partial}{\partial x}\left(D\frac{\partial C'}{\partial x}\right)} = 0,$$

$$\overline{P'} = 0,$$

and obtain for the variation in time of the temporal mean of the physical quantity C:

$$\frac{\partial\overline{C}}{\partial t} = -\frac{\partial\left(\overline{u}\,\overline{C}\right)}{\partial x} - \frac{\partial\overline{\left(u'C'\right)}}{\partial x} + \frac{\partial}{\partial x}\left(D\frac{\partial\overline{C}}{\partial x}\right) + \overline{P}. \tag{3.10}$$

We see from this that the variation in time of $\overline{C}$ indeed depends on the eddy fluctuations u' and C'; the nonlinearity of the term uC (advection flux) prevents the eddy fluctuations from being cancelled out by time averaging. From the statistical viewpoint, the quantity $\overline{u'C'} = \overline{(u-\overline{u})(C-\overline{C})}$ corresponds to the *covariance* between the quantities u and C. It vanishes if u and C are uncorrelated. From the physical viewpoint, it describes the influence of the eddy fluctuations on the temporal change of $\overline{C}$ and denotes an *eddy flux density*,

$$F = \overline{u'C'},$$

which is, unlike the molecular fluxes explained in Sect. 3.1 and described by the second term on the right-hand side of (3.5), a part of the advective flux uC. If, for example, u and C are significantly positively correlated, then a positive deviation u' goes in hand probably with a positive deviation C' and a negative deviation u' probably with a negative deviation C', whereby a transport of the quantity C in the positive direction of the x-coordinate axis results. Instead, if u and C are uncorrelated, the eddy flux density $\overline{u'C'}$ vanishes.

The motions of the eddies are seemingly stochastic, quite similar to the thermal motion of molecules. With regard to this fact we are talking about *eddy diffusion*, in contrast to the *molecular diffusion* presented in Sect. 3.1, and describe the eddy (diffusive) fluxes similar to the (molecular) diffusive fluxes. A widely used simple parameterisation assumes the eddy flux density of the physical quantity C to be proportional to the gradient of the temporal mean of C, quite similar to Fick's first law (3.1),

$$F = \overline{u'C'} = -K\frac{\partial \overline{C}}{\partial x}, \tag{3.11}$$

where K denotes the *eddy diffusion constant* (also called *eddy diffusion coefficient* or *eddy diffusivity*) with the unit $m^2 \, s^{-1}$. The latter depends, like the molecular diffusion constant D, on the physical properties of both the transporting fluid and the transported physical quantity C, but, unlike the molecular diffusion constant, furthermore, among other physical properties (for example the stability of stratification), on the air velocity field $u(x, t)$ and finally on the averaging time τ. This parameterisation takes care of the problem that the smallest eddy motions cannot be resolved by the temporal and spatial resolution of the actual climate models. In a three-dimensional isotropic space, the eddy flux density of a scalar quantity C in a flow $\vec{u}(\vec{x}, t)$ is

$$\vec{F} = \overline{\vec{u}'C'} = -K\vec{\nabla}\overline{C}, \tag{3.12}$$

in analogy to (3.2). Table 3.3 shows examples of eddy flux densities.

With this we obtain for the averaged one-dimensional advection-diffusion equation (3.10)

$$\frac{\partial \overline{C}}{\partial t} = -\frac{\partial(\overline{u}\overline{C})}{\partial x} + \frac{\partial}{\partial x}\left(K\frac{\partial \overline{C}}{\partial x}\right) + \frac{\partial}{\partial x}\left(D\frac{\partial \overline{C}}{\partial x}\right) + \overline{P} \tag{3.13}$$

Table 3.3 Examples of eddy flux densities, ρ denotes a mass density and ρ_s the particle or the mass density of salt.

transported quantity	formulation
mass	$\vec{F} = \overline{\vec{u}'\rho'} = -K\vec{\nabla}\overline{\rho}$
heat	$\vec{F} = \rho c\overline{\vec{u}'T'} = -K\rho c\vec{\nabla}\overline{T}$
salt	$\vec{F} = \overline{\vec{u}'\rho_s'} = -K\vec{\nabla}\overline{\rho}_s$
y-momentum	$\vec{F} = \rho\overline{\vec{u}'u_y'} = -\rho K\vec{\nabla}\overline{u}_y$

and analogously for the averaged three-dimensional advection-diffusion equation

$$\frac{\partial \overline{C}}{\partial t} = -\vec{\nabla} \cdot \left(\overline{\vec{u}}\,\overline{C}\right) + \vec{\nabla} \cdot \left(K\vec{\nabla}\overline{C}\right) + \vec{\nabla} \cdot \left(D\vec{\nabla}\overline{C}\right) + \overline{P}. \tag{3.14}$$

These general relations apply for the ocean, too. In the case of large-scale motions in the free atmosphere or the free ocean the molecular flux densities are mostly very small and in many cases even negligibly small compared to the eddy flux densities.

3.5 Solution of the Advection Equation

We consider the simplest case of (3.13) with a constant flow velocity u and without any diffusion, sources or sinks. This leads to the one-dimensional advection equation

$$\frac{\partial C}{\partial t} + u\frac{\partial C}{\partial x} = 0 \tag{3.15}$$

(the overbars are omitted).

3.5.1 Analytical Solution

The general solution of this equation can be written as

$$C(x,t) = f(x - ut), \tag{3.16}$$

where f is an arbitrary differentiable function. As a partial differential equation of first order in time, (3.15) requires an initial condition for $t = 0$, which is given by $f(x)$.

Equation (3.16) describes a constant movement of a concentration distribution without any changes in shape f along the positive x-axis, as illustrated in Fig. 3.5. It represents a dispersion-free propagation of a disturbance along the x-axis at constant

Fig. 3.5 Transport of function $f(x)$ along the positive x-axis in a constant flow velocity $u > 0$ under the preservation of its form.

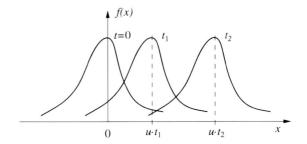

speed u and is reminiscent of a wave. Although (3.15) is not the classical wave equation, it can be shown that it is indeed part of the classical wave equation.

We note that a disturbance moving to the left is given by the following partial differential equation (PDE):

$$\frac{\partial C}{\partial t} - u\frac{\partial C}{\partial x} = 0,$$

with $u > 0$. In the following we investigate the PDE of which the solution propagates at a constant velocity along the positive as well as the negative x-axis. The following PDE satisfies these conditions:

$$\left(\frac{\partial}{\partial t} - u\frac{\partial}{\partial x}\right)\left(\frac{\partial}{\partial t} + u\frac{\partial}{\partial x}\right)C = 0.$$

The order of the operators inside the brackets may be interchanged. Eliminating the brackets and setting $u =$ constant leads to

$$\frac{\partial^2 C}{\partial t^2} - u^2 \frac{\partial^2 C}{\partial x^2} = 0. \tag{3.17}$$

This is the *classical wave equation* with a constant phase velocity u. We briefly specify the solution of the advection equation (3.15), subject to the initial condition

$$C(x,0) = Ae^{ikx}. \tag{3.18}$$

Equation (3.18) contains the function cos (real part) as well as the function sin (imaginary part). According to (3.16), a particular solution of (3.15) is therefore

$$C(x,t) = Ae^{ik(x-ut)}. \tag{3.19}$$

Equation (3.19) represents a plane wave of amplitude A. The quantities shown in Table 3.4 characterize the wave.

3.5 Solution of the Advection Equation

Table 3.4 Summary of quantities describing a one-dimensional harmonic wave.

quantity		relation
wave number	k	$k = \dfrac{2\pi}{\lambda}$
wave length	λ	$\lambda = \dfrac{2\pi}{k} = \dfrac{u}{\nu}$
angular frequency	ω	$\omega = \dfrac{2\pi}{T}$
period	T	$T = \dfrac{2\pi}{\omega} = \dfrac{1}{\nu}$
frequency	ν	$\nu = \dfrac{1}{T} = \dfrac{u}{\lambda}$

3.5.2 Numerical Solution

We now solve the one-dimensional advection equation (3.15) numerically by discretising (3.15) in space and time as follows:

Spatial discretisation: $\qquad x = m\Delta x, \qquad m = 0, 1, 2, \ldots$

Temporal discretisation: $\qquad t = n\Delta t, \qquad n = 0, 1, 2, \ldots$

We adopt the following notation

$$C(x, t) = C(m\Delta x, n\Delta t) = C_{m,n} \qquad (3.20)$$

for the values of the solution at the spatio-temporal grid points. The application of central differences in (3.15) yields

$$\frac{C_{m,n+1} - C_{m,n-1}}{2\Delta t} + u\frac{C_{m+1,n} - C_{m-1,n}}{2\Delta x} = 0. \qquad (3.21)$$

Solving for the value at the most recent time point $(n + 1)\Delta t$ yields

$$C_{m,n+1} = C_{m,n-1} - \frac{u\Delta t}{\Delta x}(C_{m+1,n} - C_{m-1,n}). \qquad (3.22)$$

This scheme is called CTCS scheme (*centered in time, centered in space*). One can see that the identification of the value of solution C at a given time requires information from two neighboring grid points of the previous time step. This is schematically illustrated on a spatio-temporal grid in Fig. 3.6. With regard to the arrangement of the "predictors" this scheme is called *leap-frog scheme*. It must be noted that for the first time step from $t = 0$ to $t = \Delta t$ the CTCS scheme does not work. Instead, we must use the Euler forward scheme for time, therefore

Fig. 3.6 Illustration of the leap-frog scheme (CTCS) on a spatio-temporal grid.

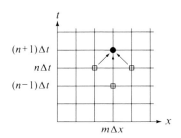

$$C_{m,1} = C_{m,0} - \frac{u\Delta t}{2\Delta x}(C_{m+1,0} - C_{m-1,0}). \quad (3.23)$$

Here we used the FTCS scheme (*forward in time, centered in space*). For $C_{m,0}$ the initial condition $C(x,0)$ is substituted.

3.5.3 Numerical Stability, CFL Criterion

The following presentation is based on Haltiner and Williams (1980). Here we explore the characteristics of the leap-frog scheme (CTCS scheme). To this end, we assume the plane wave (3.18) as initial condition. Since we know the analytical solution, we can directly derive the discretized form,

$$C_{m,n} = B^{n\Delta t} e^{ikm\Delta x}, \quad (3.24)$$

where the time dependence is given in a particular form (with an appropriate choice of B in (3.24) this is identical to (3.19)). We insert (3.24) into (3.22) and obtain

$$\left(B^{\Delta t}\right)^2 + 2i\sigma B^{\Delta t} - 1 = 0, \quad (3.25)$$

with

$$\sigma = \frac{u\Delta t}{\Delta x}\sin(k\Delta x). \quad (3.26)$$

This is a quadratic equation in $B^{\Delta t}$ with the two solutions

$$B^{\Delta t} = -i\sigma \pm \sqrt{1-\sigma^2}. \quad (3.27)$$

We distinguish two cases:

- *Stable case* $|\sigma| \leq 1$:

  Both solutions $B^{\Delta t}$ have the absolute value 1, therefore they lie on the unit circle in the complex plane (Fig. 3.7). From the figure it follows:

3.5 Solution of the Advection Equation

Fig. 3.7 Illustration of the solutions (3.27) indicated as the large dots in the complex plane.

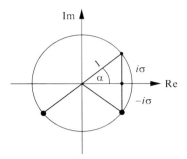

$$B^{\Delta t} = \begin{cases} e^{-i\alpha} \\ e^{i(\alpha+\pi)} \end{cases}, \qquad \sin\alpha = \sigma. \tag{3.28}$$

Therefore, the solution (3.24) can be written as

$$C_{m,n} = \left(M e^{-i\alpha n} + E e^{i(\alpha+\pi)n}\right) e^{ikm\Delta x} \tag{3.29a}$$

$$C_{m,0} = (M + E) e^{ikm\Delta x}. \tag{3.29b}$$

According to (3.18) we require $M + E = A$. Therefore, the discretised solution can be written as follows:

$$C_{m,n} = \underbrace{(A - E) e^{ik\left(m\Delta x - \frac{\alpha n}{k}\right)}}_{P} + \underbrace{(-1)^n E e^{ik\left(m\Delta x + \frac{\alpha n}{k}\right)}}_{N}, \tag{3.30}$$

where P denotes the physical mode and N the numerical mode (*computational mode*) of the solution. Note, that N changes its sign at every time step!
E remains to be identified. For the first time step we use (3.23). For the concentrations at time $t = 0$ we use (3.29b) and obtain

$$C_{m,1} = A(1 - i\sin\alpha) e^{ikm\Delta x} = (A - E) e^{ikm\Delta x - i\alpha} - E e^{ikm\Delta x + i\alpha},$$

thus

$$E = A\frac{\cos\alpha - 1}{2\cos\alpha}.$$

Inserting this expression into (3.30) yields finally

$$C_{m,n} = \underbrace{A\frac{1 + \cos\alpha}{2\cos\alpha} e^{ik\left(m\Delta x - \frac{\alpha n}{k}\right)}}_{P} + \underbrace{(-1)^{n+1} A\frac{1 - \cos\alpha}{2\cos\alpha} e^{ik\left(m\Delta x + \frac{\alpha n}{k}\right)}}_{N}.$$

$$\tag{3.31}$$

The convergence of (3.31) to (3.19) can be shown, as the following is valid:

$$\Delta x \to 0 \quad \Longrightarrow \quad \sigma = \frac{u\Delta t}{\Delta x} \sin(k\Delta x) \to uk\Delta t$$

and for $\Delta t \to 0$ it follows that $\sigma \ll 1$ and hence $\sigma = \sin\alpha \approx \alpha$. Therefore, (3.31) converges to

$$C_{m,n} \to \underbrace{A\frac{1+\cos\alpha}{2\cos\alpha}e^{ik(x-ut)}}_{P} + \underbrace{(-1)^{n+1}A\frac{1-\cos\alpha}{2\cos\alpha}e^{ik(x+ut)}}_{N}.$$

The term P describes the physical solution of a plane wave propagating to the right with an amplitude $A(1+\cos\alpha)/(2\cos\alpha)$; for $\Delta t \to 0$ the amplitude is equal to A. The term N is the *computational mode* propagating to the left with an amplitude that vanishes for $\Delta t \to 0$.

The advection equation (3.15) was solved numerically for $u = 1$, $\Delta x = 1$, and $\Delta t = 0.1$ using scheme (3.22), while (3.23) was used for the first time step. The initial condition is an amplitude of 10 at the origin, which, in the exact solution, ought to propagate to the right preserving its shape. The numerical integration shows indeed a wave package moving to the right, physically well-founded, but also the numerical mode moving to the left and changing its sign at any grid point with each time step (Fig. 3.8). Additionally, the physical mode is subject to *numerical dispersion*, meaning that its form changes. In this scheme,

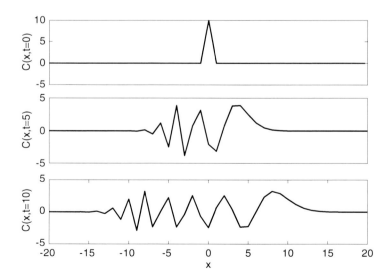

Fig. 3.8 Dissipation of a wave package and generation of the numerical mode ($x < 0$) for the solution of the advection equation (3.15) for $u = 1$, $\Delta x = 1$, and $\Delta t = 0.1$ using CTCS.

3.5 Solution of the Advection Equation

the propagation velocity of a wave depends on the wave length. This causes the initially well localized wave package to slowly disperse.

- *Unstable case* $|\sigma| > 1$:

  In this case, we can rewrite (3.27):

  $$B^{\Delta t} = -i(\sigma \pm S), \qquad S = \sqrt{\sigma^2 - 1} > 0.$$

  For $\sigma > 1$ we have $\sigma + S > 1$ and hence $|(B^{\Delta t})^n| \to \infty$ for $n \to \infty$. For $\sigma < -1$ we have $\sigma - S < -1$ and $|(B^{\Delta t})^n|$ diverges as well. The solution increases exponentially with time: it "explodes".

In consequence, the numerical solution using the CTCS scheme (3.22) only converges under the condition $|\sigma| < 1$, that is

$$\left| \frac{u \Delta t}{\Delta x} \sin(k \Delta x) \right| \leq 1.$$

For this condition to be fulfilled for all wave numbers k, the following very important condition must be satisfied:

$$\left| \frac{u \Delta t}{\Delta x} \right| \leq 1. \qquad (3.32)$$

Condition (3.32) is called the *Courant–Friedrichs–Lewy criterion* (Courant et al. 1928), which must be satisfied necessarily in order to obtain stable numerical solutions using central differences. It is usually referred to as *CFL criterion*. The CFL criterion links the velocity, at which signals are transported in the fluid, to the resolution of the space-time grid required to resolve the flow. At high transport velocities and a fixed spatial resolution, small time steps must be chosen. High flow velocities often occur in natural systems relevant for climate modelling. For example, the jet stream in the high troposphere/lower stratosphere of the mid-latitudes, or western boundary currents in ocean basins are difficult to resolve and require small time steps to satisfy the CFL criterion.

We will introduce numerical schemes which do not have to satisfy the CFL criterion and therefore are applied in difficult cases, where the time step would have to be reduced too much.

We now present a more intuitive and physical way to understand the origin of the CFL criterion. The CFL criterion is a result of the wave propagation as described in the advection equation, and the area of influence of the chosen numerical scheme.

This is illustrated on a spatio-temporal grid in Fig. 3.9. A point (x, t) on this grid is visited by a wave which started at $t = 0$ from a specific location and has propagated in time t to location x. The wave propagates along its *characteristic*; here as a special case with a constant velocity u. The characteristic of a wave is defined as the geometric location of constant phase in the space-time-continuum.

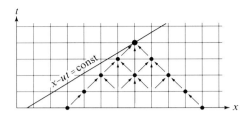

Fig. 3.9 Space-time grid and area of influence of the CTCS scheme (3.22). In the case here, the characteristic of wave propagation lies outside the area of influence and thus violates the CFL criterion.

Here, the phase is given by $\Phi = x - ut$. The CFL criterion is the requirement, that the characteristic that runs through point (x, t) is captured by the numerical scheme at all times.

The area of influence of the numerical scheme is determined by the specific formulation. In the case of the leap-frog scheme ((3.22), CTCS) a triangular area of influence in the space-time grid results. Its vertex is located at point (x, t). From Fig. 3.9 we see that the slope of the characteristic must be larger than the slope of the area of influence of the numerical scheme applied, hence

$$\frac{1}{u} \geq \frac{\Delta t}{\Delta x} \quad \Longleftrightarrow \quad \frac{u \Delta t}{\Delta x} \leq 1,$$

which yields the CFL criterion (3.32). Figure 3.9 also illustrates that the slope of area of influence decreases either by increasing Δx or by decreasing Δt as is directly evident from (3.32).

Analogously, using (3.24) for the heat equation

$$\frac{\partial T}{\partial t} = \kappa \frac{\partial^2 T}{\partial x^2}$$

and solving it numerically using the FTCS-scheme we obtain the CFL criterion

$$\frac{\kappa \Delta t}{\Delta x^2} \leq \frac{1}{2}. \tag{3.33}$$

3.6 Further Methods for the Solution of the Advection Equation

3.6.1 Euler Forward in Time, Centered in Space (FTCS)

The numerical mode in (3.30) arose from the fact that the computation of the new time step required the information of two previous steps. In order to suppress

3.6 Further Methods for the Solution of the Advection Equation

the numerical mode we try an Euler forward method for time. Hence, (3.15) in a discretized form becomes

$$C_{m,n+1} = C_{m,n} - \frac{u\Delta t}{2\Delta x}(C_{m+1,n} - C_{m-1,n}). \tag{3.34}$$

We assume

$$C_{m,n} = B^{n\Delta t} e^{ikm\Delta x} \tag{3.35}$$

and obtain

$$B^{\Delta t} = 1 - i\sigma = \sqrt{1+\sigma^2}\, e^{-i\theta}, \tag{3.36}$$

where

$$\sigma = \frac{u\Delta t}{\Delta x}\sin(k\Delta x), \qquad \tan\theta = \sigma.$$

Inserting (3.36) into (3.35) yields

$$C_{m,n} = \left(1 + \left(\frac{u\Delta t}{\Delta x}\right)^2 \sin^2(k\Delta x)\right)^{n/2} e^{ik(m\Delta x - n\theta/k)}.$$

Since the above bracket is always greater than 1, the amplitude increases with time. Therefore we find $|C_{m,n}| \to \infty$ for $n \to \infty$. The solution "explodes" using this scheme.

3.6.2 Euler Forward in Time, Upstream in Space (FTUS)

The following scheme takes into consideration the physics inherent in the simple advection equation (3.15). In a flow with speed u, the information originates from the negative x-direction and is carried at velocity u towards the grid point under consideration. It seems obvious to discretize the spatial derivative using a scheme that accounts for this situation. Instead of centered differences, Euler backwards is used. It is clearer to use the term *upstream scheme* in this context, since spatial information originating from upstream locations is used. For $u > 0$ the discretized form of (3.15) therefore becomes

$$C_{m,n+1} = C_{m,n} - \frac{u\Delta t}{\Delta x}(C_{m,n} - C_{m-1,n}). \tag{3.37}$$

Inserting (3.35) into (3.37) and simplifying, we obtain

$$B^{\Delta t} = 1 - \frac{u\Delta t}{\Delta x}\left(1 - e^{-ik\Delta x}\right). \tag{3.38}$$

The numerical scheme stays stable if $|B^{\Delta t}| \leq 1$. Based on (3.38), it can be shown that this is satisfied for all wave numbers k, provided

$$\frac{u\Delta t}{\Delta x} \leq 1, \tag{3.39}$$

hence, if the CFL criterion (3.32) is satisfied. The disadvantage of the upstream scheme is a relatively strong damping and dispersion as illustrated in Fig. 3.10. In the upstream scheme, the damping increases with the reduction of Δt.

3.6.3 Implicit Scheme

Often, the CFL criterion can only be satisfied if extremely short time steps are chosen. For example, in typical ocean models near the surface, where the isopycnal surfaces (surfaces of constant density) are steep, fluxes become large, and time steps on the order of seconds would be required to satisfy CFL. This is clearly not practical, and therefore an alternative must be found. The idea of the implicit scheme is that spatial derivatives are taken at the new time $(n + 1) \Delta t$. There are various possibilities to do so as is illustrated on a spatio-temporal grid in Fig. 3.11.

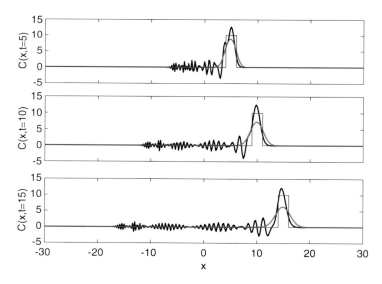

Fig. 3.10 Comparison of the exact solution (*thin blue curve*) of the advection of a rectangular profile using different numerical solutions of the advection equation: centered differences in t and x (CTCS, (3.21), *black curve*), and upstream scheme, respectively (FTUS, (3.37), *red curve*). For both, $\Delta x = 0.2$, $\Delta t = 0.1$ and $u = 1$ are used. The initial condition is $C = 1$ for $-1 \leq x \leq 1$ and $C = 0$ else. The numerical mode appearing when centered differences (3.21) are used, is obvious. The upstream scheme does not produce a numerical mode but a very strong damping and dispersion.

3.6 Further Methods for the Solution of the Advection Equation

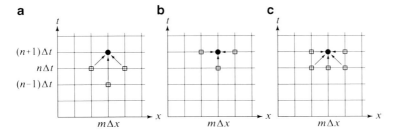

Fig. 3.11 Schematic representation of an explicit and two implicit numerical schemes. (**a**) explicit leap-frog scheme (3.21), (**b**) implicit leap-frog scheme and (**c**) implicit trapezoidal scheme (3.40).

The implementation of the implicit trapezoidal scheme for the advection equation (3.15) reads

$$\frac{C_{m,n+1} - C_{m,n}}{\Delta t} + u\frac{1}{2}\left(\frac{C_{m+1,n+1} - C_{m-1,n+1}}{2\Delta x} + \frac{C_{m+1,n} - C_{m-1,n}}{2\Delta x}\right) = 0, \tag{3.40}$$

where $\frac{1}{2}(\ldots)$ represents the average of the first spatial derivative at times $(n+1)\Delta t$ and $n\Delta t$. Again, we insert (3.35) into (3.40) and obtain

$$B^{\Delta t} = \frac{1 - i\sigma}{1 + i\sigma}, \qquad \sigma = \frac{u\Delta t}{2\Delta x}\sin(k\Delta x). \tag{3.41}$$

For any value for σ we find $|B^{\Delta t}| = 1$. Therefore, this scheme is stable without a constraint concerning the time step or the spatial grid resolution. For this scheme, neither the CFL criterion has to be satisfied nor a damping of the amplitude occurs. Unfortunately, the phase velocities of the waves become distorted.

It is evident from (3.40) that the implicit scheme leads to a large system of linear equations which requires a matrix inversion in order to solve for the new time step. We now write the equations resulting from using the implicit scheme in a compact way. Therefore, we collect the solutions at grid points $m = 1, 2, \ldots, M$ and time n in a vector:

$$\vec{C}_n = \begin{pmatrix} C_{1,n} \\ C_{2,n} \\ \vdots \\ C_{M,n} \end{pmatrix}. \tag{3.42}$$

The discretized form (3.40) can then be written as a system of linear equations in the following way:

$$\begin{pmatrix} \vdots & \vdots & \vdots & \vdots & \vdots \\ \cdots & -\frac{u\Delta t}{4\Delta x} & -1 & \frac{u\Delta t}{4\Delta x} & \cdots \\ \vdots & \vdots & \vdots & \vdots & \vdots \end{pmatrix} \begin{pmatrix} \vdots \\ C_{m-1,n} \\ C_{m,n} \\ C_{m+1,n} \\ \vdots \end{pmatrix} + \begin{pmatrix} \vdots & \vdots & \vdots & \vdots & \vdots \\ \cdots & -\frac{u\Delta t}{4\Delta x} & 1 & \frac{u\Delta t}{4\Delta x} & \cdots \\ \vdots & \vdots & \vdots & \vdots & \vdots \end{pmatrix} \begin{pmatrix} \vdots \\ C_{m-1,n+1} \\ C_{m,n+1} \\ C_{m+1,n+1} \\ \vdots \end{pmatrix} = 0,$$

or in short

$$\mathbf{A}\vec{C}_n + \mathbf{B}\vec{C}_{n+1} = 0. \tag{3.43}$$

The solution at time $n+1$ is given by

$$\vec{C}_{n+1} = -\mathbf{B}^{-1}\mathbf{A}\vec{C}_n. \tag{3.44}$$

This means that for one time step, the solution at all spatial grid points is derived by the inversion of a linear equation system. Since the corresponding matrices are usually sparse, the solution can be obtained without using a full matrix inversion which is computationally expensive. In the case of (3.40), the matrix has non-zero elements only in the diagonal and the first off-diagonals.

The numerical solution of the implicit scheme (3.40) for the same parameters Δt and Δx and the same initial conditions as in Fig. 3.10 is practically indistinguishable from the numerical solution using (3.21). However, the big advantage is the possibility of an arbitrary increase of the time step without sacrificing the quality of the numerical solution (Fig. 3.12).

3.6.4 Lax Scheme

In Sect. 3.6.1 it was shown, that the scheme Euler forward in time, centered in space (FTCS) is always unstable. Now, the idea in the Lax scheme is to stabilize the FTCS method by an additional diffusion term. This can be achieved by replacing $C_{m,n}$ by the spatial mean of two neighbouring grid points in (3.34). This leads to

$$C_{m,n+1} = \frac{1}{2}(C_{m+1,n} + C_{m-1,n}) - \frac{u\Delta t}{2\Delta x}(C_{m+1,n} - C_{m-1,n}). \tag{3.45}$$

The scheme (3.45) is equivalent to (3.34) plus a diffusive term, because

$$C_{m,n+1} = \underbrace{C_{m,n} - \frac{u\Delta t}{2\Delta x}(C_{m+1,n} - C_{m-1,n})}_{= (3.34)} + \underbrace{\frac{1}{2}(C_{m+1,n} - 2C_{m,n} + C_{m-1,n})}_{D},$$

and term D is a discretized form of a diffusion term

3.6 Further Methods for the Solution of the Advection Equation

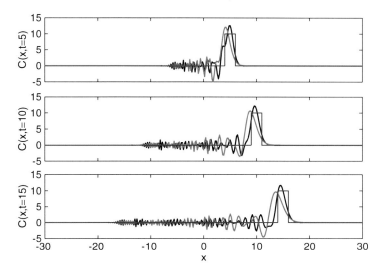

Fig. 3.12 Comparison of the exact solution (*thin blue curve*) with the numerical solutions of the advection equation using the implicit-trapezoidal scheme (3.40) with two different time steps: $\Delta t = 0.1$ (*black curve*) and $\Delta t = 0.5$ (*red curve*) with $\Delta x = 0.2$ and $u = 1$. The initial condition is $C = 1$ for $-1 \leq x \leq 1$, and $C = 0$ else. Both schemes reproduce the main maximum relatively well, but they also generate numerical modes propagating to the left. The method with the large time step exhibits a greater lag of the main maximum.

$$\text{Term D} = \frac{\Delta x^2}{2} \frac{C_{m+1,n} - 2C_{m,n} + C_{m-1,n}}{\Delta x^2} \approx \Delta t \left(\frac{\Delta x^2}{2\Delta t}\right) \frac{\partial^2 C}{\partial x^2} \quad (3.46)$$

with a numerical diffusion constant $\Delta x^2 / (2\Delta t)$. Therefore, the reduction of Δx decreases the diffusion quadratically, whereas a decrease of the time step increases diffusion. But Δx and Δt cannot be chosen independent from one another because of the CFL criterion. This follows from using form (3.35) and inserting it into (3.45). This yields

$$B^{\Delta t} = \cos(k\Delta x) - \frac{u\Delta t}{\Delta x} i \sin(k\Delta x). \quad (3.47)$$

Hence, the scheme is stable only if

$$\left|B^{\Delta t}\right| \leq 1 \quad \Longleftrightarrow \quad \left|\frac{u\Delta t}{\Delta x}\right| \leq 1, \quad (3.48)$$

which is again the classical CFL criterion. The numerical solution is illustrated in Fig. 3.13; the parameters are identical to Fig. 3.10. The smaller the chosen time step, the stronger is the effect of diffusion of the first term in (3.45) and the scheme becomes useless.

The Lax scheme exhibits no numerical mode. But the clear disadvantage of the scheme is the rather large damping of gradients.

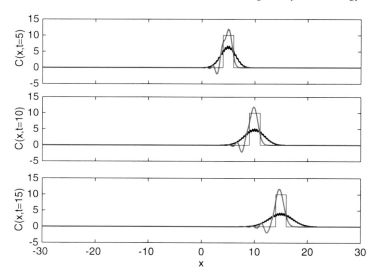

Fig. 3.13 Comparison of the exact solution (*thin blue curve*) with different numerical solutions of the advection equation: Lax scheme (3.45) (*black curve*) and Lax–Wendroff scheme (3.49) (*red curve*). The parameters are: $\Delta x = 0.2$, $\Delta t = 0.1$ and $u = 1$. The initial condition is $C = 1$ for $-1 \leq x \leq 1$, $C = 0$ else. The Lax scheme (3.45) exhibits a strong damping and therefore an underestimation of the gradients. The Lax–Wendroff scheme (3.49) overestimates the maximum and shows trailing oscillations. However, the representation of the gradients is significantly improved over the Lax scheme.

3.6.5 Lax–Wendroff Scheme

The Lax–Wendroff scheme addresses directly the problem of numerical diffusion from which the Lax scheme suffers. It reproduces gradients considerably better than the Lax scheme. This scheme is based on the idea to combine the Lax scheme for an intermediate time step with a subsequent Euler forward in time, centered differences in space (FTCS scheme, Sect. 3.6.1). The intermediate or preparatory step is given by

$$\tilde{C}_{m+\frac{1}{2},n+\frac{1}{2}} = \frac{1}{2}(C_{m+1,n} + C_{m,n}) - \frac{u\Delta t}{2\Delta x}(C_{m+1,n} - C_{m,n}) \quad (3.49)$$

and then followed by time stepping to time $(n+1)\Delta t$

$$C_{m,n+1} = C_{m,n} - \frac{u\Delta t}{\Delta x}\left(\tilde{C}_{m+\frac{1}{2},n+\frac{1}{2}} - \tilde{C}_{m-\frac{1}{2},n+\frac{1}{2}}\right). \quad (3.50)$$

Inserting (3.49) into (3.50) reveals how the formerly unstable scheme (3.34) becomes stabilized:

$$C_{m,n+1} = \underbrace{C_{m,n} - \frac{u\Delta t}{2\Delta x}(C_{m+1,n} - C_{m-1,n})}_{= (3.34)}$$
$$+ \underbrace{\frac{u^2 \Delta t^2}{2\Delta x^2}(C_{m+1,n} - 2C_{m,n} + C_{m-1,n})}_{D}. \tag{3.51}$$

Term D in (3.51) is a diffusion term, because

$$\text{Term D} = \Delta t \frac{u^2 \Delta t}{2} \frac{C_{m+1,n} - 2C_{m,n} + C_{m-1,n}}{\Delta x^2} \approx \Delta t \left(\frac{u^2 \Delta t}{2}\right) \frac{\partial^2 C}{\partial x^2}. \tag{3.52}$$

Here, the numerical diffusivity is $u^2 \Delta t/2$ and thus much weaker than for the Lax scheme. It scales with Δt, and hence decreases for small time steps. The numerical solution is illustrated in Fig. 3.13; the parameters are as in Fig. 3.10. In this scheme, the reduction of the time step does not affect the form of the main maximum but the trailing oscillations extend over a larger domain.

It can be shown that also for the Lax–Wendroff Scheme the CFL criterion (3.48) has to be satisfied to ensure stability.

3.7 Numerical Solution of the Advection-Diffusion Equation

Let us now consider the one-dimensional advection-diffusion equation (3.5) with a source term proportional to $C(x,t)$:

$$\frac{\partial C}{\partial t} = D\frac{\partial^2 C}{\partial x^2} + u\frac{\partial C}{\partial x} + bC; \tag{3.53}$$

D, u and b are constants. A generalized formulation of the discretized form of (3.53) is given by

$$\frac{C_{m,n+1} - C_{m,n}}{\Delta t} = D\frac{\theta \nabla_x^2 C_{m,n+1} + (1-\theta)\nabla_x^2 C_{m,n}}{\Delta x^2} + u\frac{\nabla_x C_{m,n}}{2\Delta x} + bC_{m,n} \tag{3.54}$$

using two centred difference operators, defined as follows:

$$\nabla_x C_{m,n} = C_{m+1,n} - C_{m-1,n},$$
$$\nabla_x^2 C_{m,n} = C_{m+1,n} - 2C_{m,n} + C_{m-1,n}. \tag{3.55}$$

θ in (3.54) is a free weighting parameter, $0 \leq \theta \leq 1$, defining the "degree of implicity" of the scheme. For $\theta = 0$ the scheme is explicit and the right-hand side of (3.54) has no time index $n+1$. The explicit scheme is stable for $D\Delta t/\Delta x^2 \leq \frac{1}{2}$.

For the parameter combination $u = 0$, $b = 0$ and $\theta = \frac{1}{2}$ (3.54) is called the *Crank–Nicholson scheme* which is absolutely stable. In general, stability of (3.54) requires

$$D\frac{\Delta t}{\Delta x^2} \leq \frac{1}{2}\frac{1}{1-2\theta} \quad \text{for } 0 \leq \theta < \frac{1}{2} \tag{3.56}$$

and for absolute stability: $\theta \geq \frac{1}{2}$.

3.8 Numerical Diffusion

Any numerical scheme exhibits non-physical properties due to the truncation. By neglecting high-order terms in the Taylor expansion, errors are introduced. Numerical diffusion is one of them, and it becomes particularly obvious when the real diffusion of physical properties needs to be quantified (e.g., mixing of tracers in a fluid system, penetration of heat into the ocean, etc.). We have encountered this already in (3.45) and (3.50), but it is also evident in Fig. 3.10 (scheme (3.37)).

In order to examine the dependence of this numerical artifact from the choice of the discretization, we look at the one-dimensional advection equation (3.15) which represents one part of the classical wave equation:

$$\frac{\partial C}{\partial t} + u\frac{\partial C}{\partial x} = 0, \tag{3.57}$$

$$\frac{\partial^2 C}{\partial t^2} - u^2\frac{\partial^2 C}{\partial x^2} = 0. \tag{3.58}$$

We discretize in space (index m) and time (index n), and write the following Taylor expansions for the spatial and time steps, respectively:

$$C_{m+1,n} = C_{m,n} + \frac{\partial C_{m,n}}{\partial x}\Delta x + \frac{1}{2!}\frac{\partial^2 C_{m,n}}{\partial x^2}\Delta x^2 + \ldots$$
$$C_{m,n+1} = C_{m,n} + \frac{\partial C_{m,n}}{\partial t}\Delta t + \frac{1}{2!}\frac{\partial^2 C_{m,n}}{\partial t^2}\Delta t^2 + \ldots . \tag{3.59}$$

In (3.59), we solve for the first derivatives and insert them into (3.57). We obtain

$$\frac{C_{m,n+1} - C_{m,n}}{\Delta t} + u\frac{C_{m+1,n} - C_{m,n}}{\Delta x} - \frac{1}{2!}\frac{\partial^2 C_{m,n}}{\partial t^2}\Delta t - u\frac{1}{2!}\frac{\partial^2 C_{m,n}}{\partial x^2}\Delta x - \ldots = 0. \tag{3.60}$$

A solution of (3.57) is also a solution of (3.58). Therefore, the second time derivative in (3.60) can be substituted using (3.58). Finally, we get

$$\frac{C_{m,n+1} - C_{m,n}}{\Delta t} + u\frac{C_{m+1,n} - C_{m,n}}{\Delta x} - \left(\frac{1}{2}u^2\Delta t + \frac{1}{2}u\Delta x\right)\frac{\partial^2 C_{m,n}}{\partial x^2} - \ldots = 0. \tag{3.61}$$

3.8 Numerical Diffusion

The third term in (3.61) is again a diffusion term. (3.61) reveals the fact, that for all 1st-order schemes consisting of the numerical formulations of derivatives, diffusion occurs. We define a *numerical diffusivity*

$$D_N = \frac{1}{2}u^2 \Delta t + \frac{1}{2}u \Delta x \tag{3.62}$$

that scales with the time and spatial steps. Various schemes exist that compensate for the numerical diffusion up to a certain point (see e.g., Smolarkiewicz 1983). Such modern schemes are denoted FCT-schemes (*flux-corrected transport*).

Chapter 4
Energy Transport in the Climate System and Its Parameterisation

4.1 Basics

In the annual mean, the Earth takes up energy between 30°S and 30°N, while it has a negative energy balance towards the poles (Fig. 4.1). Since neither a continuous warming in the lower latitudes nor a cooling in the high latitudes are observed, a strong poleward transport of energy is required. The integration of the meridional radiation balance from the South Pole to the North Pole, as it is given in Fig. 4.1, yields the heat transport, required by the radiation balance (Fig. 4.2). In each hemisphere, about $5 \cdot 10^{15}\,\mathrm{J\,s^{-1}} = 5\,\mathrm{PW}$ (Petawatt) are transported polewards. This flux is split about evenly between ocean and atmosphere. The maximum heat transport in the northern hemisphere occurs around 45°N in the atmosphere and around 20°N in the ocean. This fact points to the different mechanisms and boundary conditions (continents) responsible for the meridional heat transport. The atmosphere transports heat in a way fundamentally different from that of the ocean. The most important mechanisms are briefly explained in the following sections.

A central question is how climate models simulate heat transport and whether a certain model is able to reproduce the relevant processes of heat transport at all. It turns out that state-of-the-art three-dimensional climate models (position 3/3 in the model hierarchy of Table 2.1) simulate heat transport in the atmosphere as well as in the ocean in a physically adequate way. However, particularly models with a coarser resolution tend to underestimate the meridional heat transport in some of its important components and require unphysical corrections.

4.2 Heat Transport in the Atmosphere

The total energy per unit mass in the atmosphere is given by

$$E = \underbrace{c_V\,T}_{I} + \underbrace{g\,z}_{P} + \underbrace{L\,q}_{L} + \underbrace{\tfrac{1}{2}\left(u^2 + v^2\right)}_{K}, \tag{4.1}$$

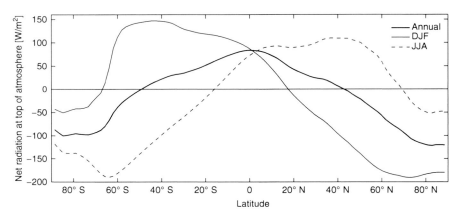

Fig. 4.1 Radiation balance as a function of latitude. Shown are the annual mean as well as the two seasonal means DJF (December-January-February) and JJA (June-July-August). Data from NCEP reanalysis (Saha et al. 2006). Figure constructed by F. Lehner.

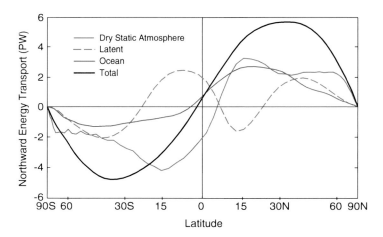

Fig. 4.2 Annual mean meridional heat transport in the atmosphere (latent and dry) and in the ocean. Figure from Siedler et al. (2001).

where c_V is the specific heat capacity of air at constant volume, T is the temperature, g is the gravity acceleration, z an altitude above a reference level, L the specific latent heat, q the humidity (mass of water vapour per mass of dry air), and u and v the horizontal components of the velocity (the vertical component is neglected). The four terms on the right-hand side denote the internal (I), the potential (P), the latent (L) and the kinetic (K) energy. The order of magnitude of the individual forms of energy in the atmosphere is given in Table 4.1.

4.2 Heat Transport in the Atmosphere

Table 4.1 Amount and distribution of energy per unit surface area in the global atmosphere (from Peixoto and Oort 1992).

		10^6 J m^{-2}	Fraction (%)
Internal energy	I	1,800	70.2
Potential energy	P	700	27.3
Latent energy	L	64	2.5
Kinetic energy	K	1.2	0.05
Total		2,565	100

In order to explain the mechanisms of the temporal and zonal mean energy flux density $\vec{F} = \vec{u}\rho E$, we split the variables into a temporal mean and a temporal deviation, on the one hand, as we have already done in Sect. 3.4, and, quite analogously, into a zonal mean and a zonal deviation, on the other hand. The temporal and zonal means of a quantity A are defined as follows:

$$\overline{A} = \frac{1}{\tau}\int_{t_1}^{t_2} A\, dt\,, \qquad [A] = \frac{1}{2\pi}\int_0^{2\pi} A\, d\lambda \qquad (4.2)$$

(time average taken over a time interval $\tau = t_2 - t_1$ of a few weeks, for example). We denote the temporal and zonal deviations from the respective means as

$$A' = A - \overline{A}\,, \qquad A^* = A - [A]\,. \qquad (4.3)$$

From (4.3) follows, that

$$\overline{A'} = 0\,, \qquad [A^*] = 0\,, \qquad (4.4)$$

as shown in (3.9).

Calculating fluxes such as the energy flux density $\vec{F} = \vec{u}\rho E$ involves products of quantities that vary in time an space. We write

$$\begin{aligned}
\overline{A\,B} &= \overline{(\overline{A} + A')(\overline{B} + B')} \\
&= \overline{\overline{A}\,\overline{B} + \overline{A}\,B' + A'\,\overline{B} + A'\,B'} \\
&= \overline{A}\,\overline{B} + \overline{A'\,B'} \\
&= \left([\overline{A}] + \overline{A}^*\right)\left([\overline{B}] + \overline{B}^*\right) + \overline{A'\,B'} \\
&= [\overline{A}][\overline{B}] + [\overline{A}]\overline{B}^* + \overline{A}^*[\overline{B}] + \overline{A}^*\overline{B}^* + \overline{A'\,B'}\,.
\end{aligned} \qquad (4.5)$$

After zonal averaging of (4.5) we obtain

$$[\overline{AB}] = [\overline{A}][\overline{B}] + 0 + 0 + \left[\overline{A}^*\,\overline{B}^*\right] + [\overline{A'B'}]$$
$$= [\overline{A}][\overline{B}] + \left[\overline{A}^*\,\overline{B}^*\right] + [\overline{A'B'}] \,. \tag{4.6}$$

The zonal and temporal mean of the product quantity AB consists of the product of the means $[\overline{A}]$ and $[\overline{B}]$ of the respective individual quantities A and B *plus* the *zonal covariance* between the temporal means $\overline{A}$ and $\overline{B}$ *plus* the zonal mean of the *temporal covariance* $\overline{A'B'}$.

For illustration, we consider the first component of (4.1) in the following. By applying (4.6) onto the meridional flux density of internal energy $v\,\rho\,c_V\,T$, where ρ is the mass density, we get, ignoring both the approximately constant mass density and the approximately constant specific heat capacity, for the zonally and temporally averaged meridional flux of internal energy:

$$[\overline{vT}] = \underbrace{[\overline{v}][\overline{T}]}_{M} + \underbrace{\left[\overline{v}^*\,\overline{T}^*\right]}_{SE} + \underbrace{[\overline{v'T'}]}_{TE} \,. \tag{4.7}$$

Hence, the zonally and temporally averaged meridional flux of internal energy consists of three components: the flux due to the mean meridional current (M), the flux due to stationary eddies (SE, caused, for example, by stationary high- and low-pressure systems) and the flux due to transient eddies (TE, caused, for example, by moving high- and low-pressure systems).

Here, M is the classical advective heat flux as described in Sect. 3.2. The terms SE and TE in (4.7) originate from spatial and temporal correlations of v and T. An illustration is given in Fig. 4.3. The meridional energy flux and its components, as determined from observations, are given in Fig. 4.4.

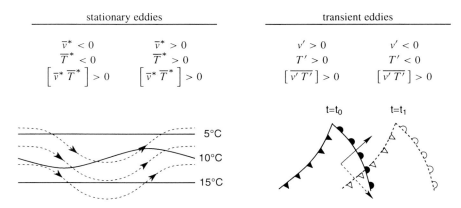

Fig. 4.3 Schematic illustration of stationary and transient eddies in the atmosphere. In the situation above, both systems transport heat northwards.

4.3 Meridional Energy Balance Model

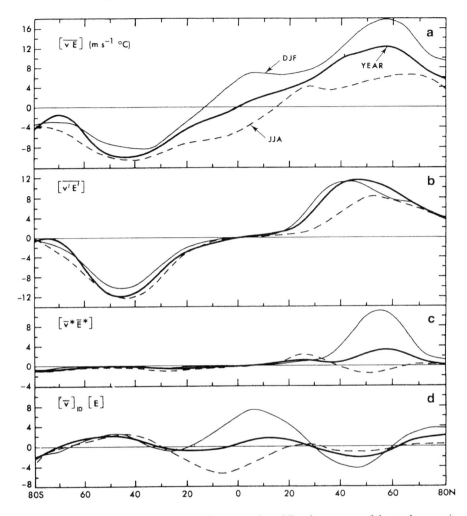

Fig. 4.4 Profile of the zonally and vertically averaged meridional transports of the total energy in (4.1) in $°C\,m\,s^{-1}$. (**a**) total; (**b**) transient eddies; (**c**) stationary eddies; (**d**) mean meridional flow, for the annual mean (*bold*), winter (*thin line*) and summer (*dashed*) months. In order to obtain units of PW, the factor $(2\pi R\cos\phi)\,c_p\,(p_0/g)$ has to be multiplied. Figure from Peixoto and Oort (1992).

4.3 Meridional Energy Balance Model

As can be inferred from Fig. 4.4, the annual mean meridional transport of total energy in the atmosphere is positive in the northern and negative in the southern hemisphere. In the zonal and annual mean, the meridional temperature gradient $\partial T/\partial \varphi$ is positive in the southern and negative in the northern hemisphere.

Therefore, a negative correlation exists between $\partial T/\partial \varphi$ and $\overline{[v E]}$. This observation-based relation is now used to suggest a simple parameterisation of the meridional heat flux. We write

$$F = \rho c \, \overline{v' T'} = -\rho c \, K(\varphi) \, \frac{1}{R} \frac{\partial T}{\partial \varphi}, \qquad (4.8)$$

where F is the meridional flux density of energy, ρ the air density, c the specific heat of air, v' and T' the eddy fluctuations of meridional air velocity and temperature, respectively. $K = K(\varphi)$ is a zonal eddy diffusivity dependent on latitude φ and on the order of 10^6 to $10^7 \, \mathrm{m^2 \, s^{-1}}$, R the Earth radius and T (the temporal mean of) the local temperature.

It is obvious, that the spatial and temporal scales, where (4.8) can be regarded as valid, are strongly limited. Figure 4.4 shows, that during winter, when steeper temperature gradients are present, more energy is transported. It has been empirically shown that (4.8) is valid for time scales of $\geq$ 6 months and spatial scales of $\geq 1,500$ km (Lorenz 1979).

We now apply this to the point energy balance model (2.1) which can be extended to a one-dimensional energy balance model. The balance equation is given by

$$h \rho c \, \frac{\partial T}{\partial t} = \frac{h}{R \cos \varphi} \frac{\partial}{\partial \varphi} \left(\rho c \, K(\varphi) \, \frac{1}{R} \frac{\partial T}{\partial \varphi} \cos \varphi \right) + \frac{1-\alpha(\varphi)}{4} S(\varphi) - \varepsilon(\varphi) \, \sigma \, T^4, \qquad (4.9)$$

where the eddy diffusivity K, the albedo α, and the emissivity ε may be functions of latitude. The (mainly short-wave) incoming radiation $S(\varphi)$ is also a function of latitude. A good approximation for the annual mean is given by

$$S(\varphi) = S_0 \left(0.5294 + 0.706 \cos^2 \varphi \right),$$

where S_0 is the solar constant.

The first term on the right-hand side of (4.9) is the divergence of the meridional heat flux density multiplied by h, the vertical extent of the troposphere. The temperature is a function of time and latitude. Since (4.9) is a differential equation of 2nd order ($\partial^2/\partial \varphi^2$) in space, two boundary conditions must be satisfied. The boundary conditions at the two poles require the heat flux to vanish, hence

$$\frac{\partial T}{\partial \varphi} = 0 \qquad \text{for} \qquad \varphi = -\frac{\pi}{2}, +\frac{\pi}{2}. \qquad (4.10)$$

The one-dimensional energy balance model presented in (4.9) is referred to as the *Budyko–Sellers EBM*. Budyko (1969) and Sellers (1969) were the first to propose such a simplified climate model and to address fundamental questions concerning climate change using their models.

The EBM in (4.9) can be further generalized to two dimensions by additionally considering the zonal direction. Such models were developed in the 1980s for studying the temperature difference between glacial and interglacial periods based on the changes in the radiation balance (North et al. 1983). Still today, they are implemented in some models of reduced complexity (Table 2.1, dimensions 2/2 and 2/3).

It must be emphasized that dynamic global circulation models of the atmosphere (AGCMs) compute the individual contributions to the energy transport (see (4.1) and (4.7)) based on the dynamics, and, to describe large-scale eddies and their effect on the heat transport, simplified parameterisations like (4.8) are not needed. This requires a minimum spatial resolution of 1,000 km or less in order to simulate eddies and their transport. As a result, a significantly increased computational burden is carried which in turn limits the length of the integrations and hence the applicability of GCMs.

4.4 Heat Transport in the Ocean

The meridional heat transport in the ocean is caused by completely different mechanisms from those operating in the atmosphere, even though the equations describing the flow are analogous in both systems. The reason for this is on the one hand, that parameters in these equations are different (in certain cases by orders of magnitude) and that on the other hand, the ocean is restrained by basin boundaries. Along the latter, important current systems emerge which contribute significantly to the meridional heat flux.

In the ocean, eddies appear to play a minor role for the meridional heat transport except in some particular regions (equator, circumpolar current, southern tip of Africa). However, this statement is based on idealized model simulations and sparse observational data, for which reason the uncertainties are still quite high. An estimate for the meridional heat transport in the global ocean is given in Fig. 4.5. Some 2 PW are transported polewards in both hemispheres, with the maximum in the northern hemisphere located more towards the equator than in the southern hemisphere.

The different ocean basins transport different amounts of heat in the different basins. Ganachaud and Wunsch (2000) roughly derived the heat fluxes based from temperature and salinity measurements combined with inverse modelling. This is illustrated in Fig. 4.6. While heat is transported northwards at all latitudes in the Atlantic, a southward transport can be observed in the Indian Ocean. Despite its large extent, the transports in the Pacific are surprisingly small. Transport in the circumpolar current is largest with about $1.3 - 1.7$ PW eastwards. The direction of the heat transport in the different ocean basins is qualitatively consistent with the strongly simplified depiction of the global oceanic conveyor belt proposed by Wally Broecker (1987).

In order to quantify the transport mechanisms of heat in the ocean, we define the vertical averaging of quantity A in the ocean according to

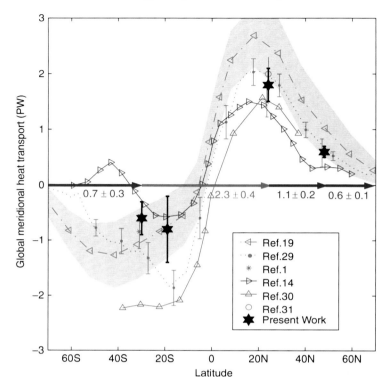

Fig. 4.5 Zonally integrated heat transport in the ocean based on observations and inverse modelling. Figure from Ganachaud and Wunsch (2000).

$$\overline{A} = \frac{1}{H} \int_{-H}^{0} A \, dz \,, \qquad A' = A - \overline{A} \,, \qquad (4.11)$$

and obtain, analogously to (4.7), the following partitioning of the temporal and zonal averaged meridional heat transport:

$$\left[\,\overline{vT}\,\right] = \underbrace{[\overline{v}][\overline{T}]}_{=0} + \underbrace{\left[\overline{v}^*\,\overline{T}^*\right]}_{G} + \underbrace{\overline{[v'][T']}}_{MOC} + \underbrace{\overline{[v'^* T'^*]}}_{EK} - \underbrace{\frac{K}{R}\left[\frac{\partial T}{\partial \varphi}\right]}_{D}, \qquad (4.12)$$

where the first term vanishes due to mass conservation in a closed basin, G denotes the heat transport associated with horizontal barotropic *gyres* (i.e., ocean gyres with a one-to-one correspondence of density and pressure, so that isobaric surfaces are isopycnic surfaces, as further explained in Sect. 6.8), MOC is the *meridional overturning circulation* (thermohaline and wind-driven meridional ocean circulation)

4.4 Heat Transport in the Ocean

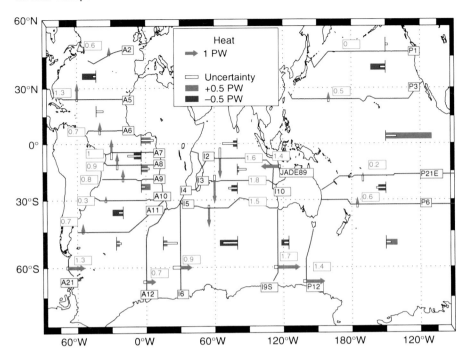

Fig. 4.6 Meridional and vertical heat transports in the different regions of the world ocean. Numbers in boxes denote the meridional transport in PW. Horizontal bars represent the vertical transport (to the left = downwards). Figure from Ganachaud and Wunsch (2000).

and EK is the heat transport due to the surface- and bottom Ekman circulation (which is induced by pressure forces, wind- and bottom-friction forces as well as Coriolis forces, see Sects. 6.2 and 6.7). The term D follows from (4.8) and is primarily important in ocean models of coarse resolution, containing eddy diffusivity.

Available data for the ocean does not yet permit to determine (4.12) by measurements. Therefore, Bryan (1987) simulated (4.12) in an ocean model of coarse resolution without eddies and found that around 80% of the meridional heat transport in the Atlantic is caused by the MOC. These results were later corroborated by a global OGCM of high resolution (Jayne and Marotzke 2002). Thanks to a resolution of 0.25°, this model simulates individual eddies.

Globally, as well as in the Atlantic, the meridional transport of heat is predominantly associated with the term MOC in (4.12). Eddies only contribute in some limited regions to the total heat transport mainly in the tropical Pacific and in the western boundary currents (Fig. 4.7). For this reason, particularly in the Atlantic, the deep circulation, or *thermohaline circulation* (which is driven by ocean water density differences emerging from temperature and salinity differences), is the most relevant one for climate.

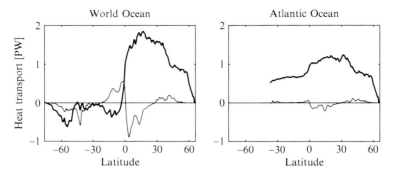

Fig. 4.7 Zonally integrated meridional heat transport for the whole ocean and the Atlantic, total flux (*bold*) and contributions by eddies (*thin lines*), simulated with a high-resolution OGCM. Figure from Jayne and Marotzke (2002).

Table 4.2 Components of eddy fluxes, namely x-, y- and z-components of eddy fluxes of horizontal momentum, eddy fluxes of heat and eddy fluxes of salt.

$-A_H \frac{\partial u}{\partial x}$, $\quad -A_H \frac{\partial u}{\partial y}$	$-A_H \frac{\partial v}{\partial x}$, $\quad -A_H \frac{\partial v}{\partial y}$	eddy-momentum flux in x- and y-direction
$-A_V \frac{\partial u}{\partial z}$	$-A_V \frac{\partial v}{\partial z}$	eddy-momentum flux in z-direction
$-K_H \frac{\partial T}{\partial x}$, $\quad -K_H \frac{\partial T}{\partial y}$	$-K_H \frac{\partial S}{\partial x}$, $\quad -K_H \frac{\partial S}{\partial y}$	eddy-heat and -salt flux in x- and y-direction
$-K_V \frac{\partial T}{\partial z}$	$-K_V \frac{\partial S}{\partial z}$	eddy-heat and -salt flux in z-direction

A rough estimate of the quantity of the term MOC in (4.12) yields the following values: In the northern Atlantic, the thermohaline circulation transports some $20 \cdot 10^6 \, \text{m}^3 \, \text{s}^{-1}$ polewards near the surface at around 18°N. Meanwhile the same volume flows towards the equator at a depth of 2–3 km along the western boundary at a temperature of around 3°C. This corresponds to a meridional heat transport of $\rho c \, (\Delta V / \Delta t) \, \Delta T \approx 10^3 \cdot 4 \cdot 10^3 \cdot 20 \cdot 10^6 \cdot 15 \, \text{W} = 1.2 \, \text{PW}$. This is in rough agreement with the values of Fig. 4.6. The large vertical temperature contrast is therefore the reason for the meridional heat transport in the Atlantic.

Also in ocean models, sub-scale transports need to be parametrised due to the limitations imposed by the grid resolution. To this end, like in the energy balance model (4.9), a flux-gradient relationship (see (4.12), term D) is chosen, because there are physical mechanisms (barotropic and baroclinic instabilities, see Pedlosky (1987)) that scale with the gradients of temperature and velocity. Therefore, the assumptions shown in Table 4.2 are made.

The values of the eddy viscosities A_H, A_V and the eddy diffusivities K_H, K_V are insufficiently restrained by data and hence, they are very uncertain. The value of A_H

4.4 Heat Transport in the Ocean

Table 4.3 Values for eddy viscosities and eddy diffusivities in ocean models of coarse resolution.

	Typical values ($m^2\ s^{-1}$)
A_H	$10^1 \ldots 10^5$
A_V	$10^{-5} \ldots 10^{-1}$
K_H	$10^3 \ldots 10^4$
K_V	$10^{-5} \ldots 10^{-4}$

depends on the grid resolution of the ocean model: the smaller Δx, the smaller A_H, since the model is able to resolve more scales for smaller Δx. Table 4.3 lists typical values used in ocean models.

The role of eddies in mixing the water masses and their realistic and consistent parameterisation in models is a current topic of research. In which way the mixing effect of the tides and their interaction with the ocean topography could be accounted for, also remains an unresolved question.

Chapter 5
Initial Value and Boundary Value Problems

5.1 Basics

The energy balance models by Sellers (1969) and Budyko (1969) result in a linear partial differential equation of 1st order in time and 2nd order in space, (4.9). The first term on the right-hand side is the divergence of the temperature gradient in one dimension, the second term is a source term, independent from the solution itself, and finally there is a term proportional to T^4 which in its linear approximation about the temperature T_o reads

$$T^4 \approx T_o^4 + \left.\frac{d\left(T^4\right)}{dT}\right|_{T_o} (T - T_o) = T_o^4 + 4\,T_o^3\,(T - T_o) = -3\,T_o^4 + 4\,T_o^3\,T.$$

If the eddy diffusivity K in (4.9) is taken as a constant, (4.9) is therefore approximately of the general type

$$\frac{\partial C}{\partial t} + K\,\vec{\nabla}^2 C + \tilde{\alpha}\,C = \tilde{\rho}(\vec{x})\,, \qquad (5.1)$$

where $\tilde{\alpha}$ is constant and $\tilde{\rho}(\vec{x})$ a function of $\vec{x}$, and defined on a (not necessarily finite) domain Ω. It describes numerous linear or linearized phenomena in physics, chemistry or mathematical biology.

Functions $C = C(\vec{x}, t)$, $\vec{x} \in \Omega$, solve (5.1) for suitable boundary and initial conditions. If such a solution results with an *initial condition*

$$C(\vec{x}, 0) = f(\vec{x})\,, \qquad (5.2)$$

where $f(\vec{x})$ is a suitable function defined on the domain Ω, then the differential equation (5.1) and the initial condition (5.2) together represent an *initial value problem*. Instead, if the problem is independent of time,

$$\vec{\nabla}^2 C + \alpha\,C = \rho(\vec{x}) \qquad (5.3)$$

T. Stocker, *Introduction to Climate Modelling*, Advances in Geophysical and Environmental Mechanics and Mathematics, DOI 10.1007/978-3-642-00773-6_5,
© Springer-Verlag Berlin Heidelberg 2011

(with the constant α and the function $\rho(\vec{x})$), for example

$$\vec{\nabla}^2 C = 0 \qquad \text{Laplace Equation,}$$
$$\vec{\nabla}^2 C = \rho(\vec{x}) \qquad \text{Poisson Equation,}$$
$$\vec{\nabla}^2 C + \alpha C = 0 \qquad \text{Helmholtz Equation,}$$

and the solution $C = C(\vec{x})$ results with *boundary conditions*

$$\alpha(\vec{x}_b) \frac{\partial C(\vec{x}_b)}{\partial n} + \beta(\vec{x}_b) C(\vec{x}_b) = \gamma(\vec{x}_b), \tag{5.4}$$

where $\partial/\partial n$ is the derivative perpendicular to the boundary $\vec{x}_b$ of the domain Ω, and $\alpha(\vec{x}_b)$, $\beta(\vec{x}_b)$ as well as $\gamma(\vec{x}_b)$ are suitable functions defined on this boundary, then the differential equation (5.3) and the boundary condition (5.4) together constitute a *boundary value problem*. For boundary conditions (5.4) at a point $\vec{x}_b$ on the boundary the following names are commonly used:

$$\alpha(\vec{x}_b) = 0 \qquad \text{Dirichlet boundary condition,}$$
$$\beta(\vec{x}_b) = 0 \qquad \text{Neumann boundary condition,}$$
$$\text{else} \qquad \text{Cauchy boundary condition.}$$

One of the most common boundary value problem is Poisson's Equation

$$\vec{\nabla}^2 C = \rho(\vec{x}) \tag{5.5}$$

(together with a suitable boundary condition), i.e., specifically in two dimensions using Cartesian coordinates,

$$\frac{\partial^2 C}{\partial x^2} + \frac{\partial^2 C}{\partial y^2} = \rho(x, y). \tag{5.6}$$

Equations (5.5) and (5.6) describe such diverse examples as stationary temperature distributions (T instead of C) in regions where heat sources are present, also stationary distributions of the electrostatic potential (φ instead of C) in regions containing electric charges, or the stationary flow of an incompressible and inviscid fluid (velocity potential instead of C) in the presence of mass sources and sinks.

5.2 Direct Numerical Solution of Poisson's Equation

This section is given only for introductory purposes and in order to demonstrate the principles. The numerical solution of a boundary value problem would not be derived by means of this method, because it would be rather inefficient to find

5.2 Direct Numerical Solution of Poisson's Equation

Fig. 5.1 Grid with inner points (circles) and boundary points (diamonds).

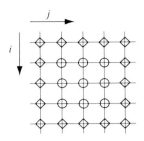

an approximate solution. Superior methods are available which will be presented below.

For simplification, we first assume, that the region, in which the equation is to be solved, is quadratic in shape. For the numerical solution of (5.6) a grid with a grid spacing of Δx and Δy (Fig. 5.1) is overlaid on the region. Circles are termed *inner points*, diamonds denote *boundary points*. Further, we assume that Dirichlet boundary conditions are formulated, i.e. the values of the boundary points are given. The derivatives in (5.6) are discretized according to Table 2.2:

$$\frac{\partial^2 C}{\partial x^2} = \frac{C_{i,j+1} - 2\,C_{i,j} + C_{i,j-1}}{\Delta x^2} + O(\Delta x^2) \qquad (5.7a)$$

$$\frac{\partial^2 C}{\partial y^2} = \frac{C_{i+1,j} - 2\,C_{i,j} + C_{i-1,j}}{\Delta y^2} + O(\Delta y^2). \qquad (5.7b)$$

Inserting (5.7) into (5.6), assuming $\Delta x = \Delta y$, and neglecting terms of higher order in (5.7), we obtain:

$$C_{i+1,j} + C_{i-1,j} + C_{i,j+1} + C_{i,j-1} - 4\,C_{i,j} = \Delta x^2 \, \rho_{i,j}. \qquad (5.8)$$

Equation (5.8) states, that the deviation of the sum of the four closest neighbors from the value in the centre is equal to the source term at this point. Equation (5.8) is a system of linear equations of dimension $NM \times NM$ of the unknowns $C_{i,j}$, $i = 1, \ldots, N$; $j = 1, \ldots, M$. By appropriately numbering the indices of the inner points, a vector $C_k, k = 1, \ldots, NM$ can be defined. We choose the following numbering, here illustrated for $N = 3$ and $M = 3$, a total of 9 inner points as in Fig. 5.1,

$$\begin{pmatrix} C_{1,1} & C_{1,2} & C_{1,3} \\ C_{2,1} & C_{2,2} & C_{2,3} \\ C_{3,1} & C_{3,2} & C_{3,3} \end{pmatrix} \equiv \begin{pmatrix} C_1 & C_4 & C_7 \\ C_2 & C_5 & C_8 \\ C_3 & C_6 & C_9 \end{pmatrix}, \qquad (5.9)$$

which converts (5.8) into the system of linear equations

$$\begin{pmatrix} -4 & 1 & 0 & 1 & 0 & 0 & 0 & 0 & 0 \\ 1 & -4 & 1 & 0 & 1 & 0 & 0 & 0 & 0 \\ 0 & 1 & -4 & 0 & 0 & 1 & 0 & 0 & 0 \\ 1 & 0 & 0 & -4 & 1 & 0 & 1 & 0 & 0 \\ 0 & 1 & 0 & 1 & -4 & 1 & 0 & 1 & 0 \\ 0 & 0 & 1 & 0 & 1 & -4 & 0 & 0 & 1 \\ 0 & 0 & 0 & 1 & 0 & 0 & -4 & 1 & 0 \\ 0 & 0 & 0 & 0 & 1 & 0 & 1 & -4 & 1 \\ 0 & 0 & 0 & 0 & 0 & 1 & 0 & 1 & -4 \end{pmatrix} \begin{pmatrix} C_1 \\ C_2 \\ C_3 \\ C_4 \\ C_5 \\ C_6 \\ C_7 \\ C_8 \\ C_9 \end{pmatrix} = \begin{pmatrix} r_1 \\ r_2 \\ r_3 \\ r_4 \\ r_5 \\ r_6 \\ r_7 \\ r_8 \\ r_9 \end{pmatrix}, \qquad (5.10)$$

where the vector $\vec{r}$ contains the values $\Delta x^2 \, \rho_{i,j}$ plus possible boundary values. The matrix in (5.10) is symmetric and has a block structure. By inverting the matrix in (5.10), C can easily be solved for. However, this method quickly leads to very large systems, which can hardly be handled. By numbering (5.9) in a different way, we obtain a different structure of the matrix. The conditioning of the matrix depends on this numbering. This has an impact on the accuracy of the solution C.

We have seen that the numerical solution of partial differential equations rapidly leads to large systems of linear equations which have to be solved using appropriate numerical methods. For a typical grid resolution of 50×50 already a matrix of dimension 2500×2500 has to be inverted.

5.3 Iterative Methods

The inversion of a large matrix is costly. To avoid this obstacle, we consider here iterative methods, first methods of *relaxation* and then the method of *successive overrelaxation*.

5.3.1 Methods of Relaxation

The solution of (5.6) is a special solution of the time-dependent partial differential equation

$$\frac{1}{K} \frac{\partial C}{\partial t} = \frac{\partial^2 C}{\partial x^2} + \frac{\partial^2 C}{\partial y^2} - \rho(\vec{x}), \qquad (5.11)$$

namely the one for which $\partial C/\partial t = 0$. We seek the stationary solution of (5.11). Discretization in space and time yields

$$C_{i,j}^{n+1} = C_{i,j}^n + \frac{K \Delta t}{\Delta x^2} \left(C_{i+1,j}^n + C_{i-1,j}^n + C_{i,j+1}^n + C_{i,j-1}^n - 4 C_{i,j}^n \right) - K \Delta t \, \rho_{i,j}, \qquad (5.12)$$

where again $\Delta x = \Delta y$ and the upper index n denotes the time step. For the time discretization in (5.12), Euler forward was used. The simultaneous solution of a system of linear equations is replaced by an iterative calculation rule given by (5.12).

5.3 Iterative Methods

In the course of a relaxation iteration procedure the values $C_{i,j}$ converge to the values of the stationary solution $\partial C/\partial t = 0$. For the solution to be stable, the appropriate CFL criterion (3.33) in two dimensions must be satisfied, i.e.

$$\frac{K \Delta t}{\Delta x^2} \le \frac{1}{4}. \tag{5.13}$$

By considering the maximum allowable time step derived from (5.13), (5.12) transforms to the classical *Jacobi method*:

$$C_{i,j}^{n+1} = \frac{1}{4}\left(C_{i+1,j}^n + C_{i-1,j}^n + C_{i,j+1}^n + C_{i,j-1}^n\right) - \frac{\Delta x^2}{4} \rho_{i,j}. \tag{5.14}$$

The Jacobi method converges only very slowly. A related method is the *Gauss–Seidel method*, which uses already computed values of the consecutive time steps in (5.14). Hence, when we proceed along the rows (i = constant) from small to large j, (5.14) can be modified to

$$C_{i,j}^{n+1} = \frac{1}{4}\left(C_{i+1,j}^n + C_{i-1,j}^{n+1} + C_{i,j+1}^n + C_{i,j-1}^{n+1}\right) - \frac{\Delta x^2}{4} \rho_{i,j}. \tag{5.15}$$

Even the Gauss–Seidel method is not very efficient. In order to reduce the error of the solution by p orders of magnitude, i.e., by a factor of 10^p, about $\frac{1}{2} p J^2$ iterations are required, where J is the number of grid points.

5.3.2 Method of Successive Overrelaxation (SOR)

Until 1970, the method of successive overrelaxation (SOR) was the standard algorithm to solve boundary value problems. For simple problems that do not have to be designed for efficiency, the SOR method is still a good and appropriate method. The SOR method is an iterative method based on the discretization given in (5.8).

The solution matrix C in (5.9) is again numbered as a vector: $C_k, k = 1, \ldots, J$, $J = MN$. For clarity, the solution vector here will be denoted x, instead of C. Hence, (5.8) reads

$$\mathbf{A}x = b. \tag{5.16}$$

The matrix $\mathbf{A}$ can be written as a sum of the diagonal, a left and a right triangular matrix

$$\mathbf{A} = \mathbf{D} + \mathbf{L} + \mathbf{R}. \tag{5.17}$$

In this notation, the methods we have previously presented read:

$$\begin{aligned}\text{\textit{Jacobi method}} \qquad & \mathbf{D}\,x^{n+1} = -(\mathbf{L}+\mathbf{R})\,x^n + b, & (5.18)\\ \text{\textit{Gauss–Seidel method}} \qquad & (\mathbf{D}+\mathbf{L})\,x^{n+1} = -\mathbf{R}\,x^n + b. & (5.19)\end{aligned}$$

We subtract $(\mathbf{D} + \mathbf{L})\, x^n$ from both sides of (5.19) and solve for x^{n+1}. We get the following equation

$$x^{n+1} = x^n - (\mathbf{D} + \mathbf{L})^{-1} \underbrace{\left((\mathbf{D} + \mathbf{L} + \mathbf{R})\, x^n - b \right)}_{=\xi^n}. \tag{5.20}$$

The quantity ξ^n is called the residual of (5.20) at time step n, because $\xi^n = \mathbf{A}\, x^n - b$. Hence, the iteration reads

$$x^{n+1} = x^n - \underbrace{(\mathbf{D} + \mathbf{L})^{-1}\, \xi^n}_{\text{COR}}, \tag{5.21}$$

where COR is the correction of the estimate for the solution at the nth iteration.

The idea of the method of successive overrelaxation is to accelerate the convergence by scaling the correction in (5.21) by a factor ω with $1 < \omega < 2$. This amounts to increasing the correction term by up to 100%. Accordingly, the SOR method reads

$$x^{n+1} = x^n - \omega\, (\mathbf{D} + \mathbf{L})^{-1}\, \xi^n \tag{5.22}$$

($\omega = 1$ would lead to the Gauss–Seidel method).

It can be shown that in order to reduce the error by a factor 10^p, here only $\frac{1}{3}\, p\, J$ iterations are required. The computational burden therefore only scales linearly with J rather than quadratically ($\frac{1}{2}\, p\, J^2$) as for the Jacobi and the Gauss–Seidel methods. However, this only holds if an optimum value for ω is used in (5.22) and this is the difficulty in the SOR method. Luckily, there are some prior estimates for ω_{opt} (see Press et al. 1992, chapter Relaxation Methods). For smaller problems, ω_{opt} can be found by a search algorithm.

The matrix formulation (5.22) of the algorithm is only of theoretical value. The practical implementation is straightforward. The discrete form of a partial differential equation of second order can be written in a generalized way as follows:

$$a_{i,j}\, x_{i+1,j} + b_{i,j}\, x_{i-1,j} + c_{i,j}\, x_{i,j+1} + d_{i,j}\, x_{i,j-1} + e_{i,j}\, x_{i,j} = f_{i,j}. \tag{5.23}$$

The new estimate for $x_{i,j}$ can be calculated analogously to (5.22):

$$x_{i,j}^{n+1} = x_{i,j}^n - \omega\, \frac{\zeta_{i,j}^n}{e_{i,j}}, \tag{5.24}$$

where $\zeta_{i,j}^n$ is the residual of the nth iteration:

$$\zeta_{i,j}^n = a_{i,j}\, x_{i+1,j}^n + b_{i,j}\, x_{i-1,j}^n + c_{i,j}\, x_{i,j+1}^n + d_{i,j}\, x_{i,j-1}^n + e_{i,j}\, x_{i,j}^n - f_{i,j}. \tag{5.25}$$

The use of previously computed $x_{i,j}^{n+1}$ in (5.25) accelerates the procedure.

Chapter 6
Large-Scale Circulation in the Ocean

Every fluid parcel in the atmosphere and the ocean obeys the fundamental laws of fluid mechanics including the equation of motion and the continuity equation. In the following we will describe approximate forms of these two equations for large-scale circulations in the ocean. Analogous equations apply for large-scale circulations in the atmosphere, too. As a preparatory step, we consider a special time derivative.

6.1 Material Derivative

Given a small water parcel moving through the ocean on a path

$$\vec{r}(t) = \begin{pmatrix} x(t) \\ y(t) \\ z(t) \end{pmatrix}.$$

Hence, at time t the water parcel passes the coordinates $x(t)$, $y(t)$ and $z(t)$ with the velocity

$$\vec{u}(t) = \frac{\mathrm{d}\vec{r}(t)}{\mathrm{d}t} = \begin{pmatrix} \frac{\mathrm{d}x(t)}{\mathrm{d}t} \\ \frac{\mathrm{d}y(t)}{\mathrm{d}t} \\ \frac{\mathrm{d}z(t)}{\mathrm{d}t} \end{pmatrix} = \begin{pmatrix} u(t) \\ v(t) \\ w(t) \end{pmatrix}, \quad (6.1)$$

where $u(t)$, $v(t)$ and $w(t)$ are the x-, y- and z-components of the velocity, respectively.

Any physical property A of the water parcel – such as the velocity, the pressure, the density, the temperature, or the salinity – is a function of time and space, $A = A(t, x, y, z)$. The total derivative with respect to time of this mathematical function is

$$\frac{\mathrm{d}A}{\mathrm{d}t} = \frac{\partial A}{\partial t} + \frac{\mathrm{d}x}{\mathrm{d}t}\frac{\partial A}{\partial x} + \frac{\mathrm{d}y}{\mathrm{d}t}\frac{\partial A}{\partial y} + \frac{\mathrm{d}z}{\mathrm{d}t}\frac{\partial A}{\partial z}. \quad (6.2)$$

Determining the derivative along the path of the water parcel, where $\mathrm{d}x/\mathrm{d}t = u$, $\mathrm{d}y/\mathrm{d}t = v$ and $\mathrm{d}z/\mathrm{d}t = w$ according to (6.1), we get the *material derivative*, also called Lagrangian derivative or advective derivative,

$$\frac{DA}{Dt} = \frac{\partial A}{\partial t} + u\frac{\partial A}{\partial x} + v\frac{\partial A}{\partial y} + w\frac{\partial A}{\partial z}, \tag{6.3}$$

corresponding to the time rate of change of the physical quantity A measured by an observer moving with the water parcel. The first term on the right-hand side of this equation is the partial derivative of A with respect to time t (the space coordinates x, y and z are held constant), called *Eulerian derivative*, corresponding to the rate of change of the physical quantity A measured by an observer at a fixed position in space (x, y, z = constant). The difference between the material derivative DA/Dt and the Eulerian derivative $\partial A/\partial t$ is due to transport with the oceanic current, of which the water parcel is part, namely due to advection. Equation (6.3) can also be written in vector notation:

$$\begin{aligned}\frac{DA}{Dt} &= \frac{\partial A}{\partial t} + (u, v, w) \cdot \begin{pmatrix} \frac{\partial A}{\partial x} \\ \frac{\partial A}{\partial y} \\ \frac{\partial A}{\partial z} \end{pmatrix} \\ &= \frac{\partial A}{\partial t} + \vec{u} \cdot \vec{\nabla} A.\end{aligned} \tag{6.4}$$

6.2 Equation of Motion

The small water parcel satisfies the conservation equation for momentum, namely the equation of motion based on the 2nd Law of Newton. With respect to an Earth-fixed coordinate system it is

$$\frac{D\vec{u}}{Dt} = \vec{a} + \vec{a}_\mathrm{I}, \tag{6.5}$$

where $\vec{a}$ signifies the acceleration (force per unit mass) due to the resultant of all *real forces* (pressure gradient force, friction force, gravity force) and $\vec{a}_\mathrm{I}$ analogously the acceleration due to the resultant of all *inertial forces* (also called *apparent forces*) arising from Earth's rotation in an Earth-fixed reference system (Coriolis force, centrifugal force).

Denoting the nearly constant angular velocity of Earth's rotation with $\vec{\Omega}$, the acceleration of the water parcel due to inertial forces relative to an Earth-fixed reference system is given by

$$\vec{a}_\mathrm{I} = -2\,\vec{\Omega} \times \vec{u} - \vec{\Omega} \times \left(\vec{\Omega} \times \vec{r}\right), \tag{6.6}$$

as shown for example in Peixoto and Oort (1992). The first term on the right-hand side is the Coriolis acceleration, the second term is the centrifugal acceleration. Due to the centrifugal acceleration, the Earth surface is approximately a rotational

6.2 Equation of Motion

Fig. 6.1 Local Cartesian coordinate system on a rotating sphere.

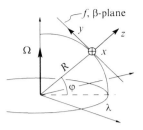

ellipsoid so that the horizontal component of the centrifugal acceleration is nearly cancelled out by the horizontal component of the gravity acceleration $\vec{g}$. Therefore, in contrast to the Coriolis force, the horizontal component of the centrifugal force can be neglected.

Every realistic version of the equation of motion has to consider that the oceanic currents are flowing on the approximately spherical surface of the Earth. But there are good approximations, especially for circulations on smaller scales, which assume oceanic flows to occur in a plane (Fig. 6.1). This plane is called f-plane or β-plane, depending on the approximations assumed (see below). A Cartesian coordinate system (x, y, z), in which the equations are formulated, is defined on this plane. The coordinate system is attached to the sphere and rotates with it. The Coriolis acceleration in this system is given by

$$\vec{a}_C = -2 \begin{pmatrix} 0 \\ \Omega \cos\varphi \\ \Omega \sin\varphi \end{pmatrix} \times \begin{pmatrix} u \\ v \\ 0 \end{pmatrix} = \begin{pmatrix} 2\Omega \sin\varphi\, v \\ -2\Omega \sin\varphi\, u \\ 2\Omega \cos\varphi\, u \end{pmatrix}; \tag{6.7}$$

considering inertial forces, we will neglect vertical motions by setting $w = 0$. In both horizontal components the common factor

$$f = 2\Omega \sin\varphi \tag{6.8}$$

appears. This is the *Coriolis parameter* which, due to the spherical shape of the Earth, depends on latitude φ. Linearisation of $f(\varphi)$ yields

$$\begin{aligned} f(\varphi) &\approx f(\varphi_0) + \left.\frac{df}{d\varphi}\right|_{\varphi_0} (\varphi - \varphi_0) \\ &\approx 2\Omega \sin\varphi_0 + 2\Omega \cos\varphi_0\, (\varphi - \varphi_0) \\ &\approx f_0 + \frac{2\Omega \cos\varphi_0}{R}\, y \\ &\approx f_0 + \beta\, y. \end{aligned} \tag{6.9}$$

If we only account for the constant term f_0 in the equations of motion in the (x, y, z) system, we are considering the dynamics on an f-plane. On the β-plane one uses the linear approximation (6.9) when considering the dynamics.

Next, we work out the most important real forces, namely the pressure gradient force, the friction force due to shear stress and the gravity force. Pressure forces (caused by pressure p) and friction forces (caused by shear stress τ) act on a mass element as follows (see also Fig. 6.2):

$$\begin{aligned}\rho\,\delta x\,\delta y\,\delta z\,a_x = {}& p(x)\,\delta y\,\delta z - p(x+\delta x)\,\delta y\,\delta z \\ & + \tau_{xy}(y+\delta y)\,\delta z\,\delta x - \tau_{xy}(y)\,\delta z\,\delta x \\ & + \tau_{xz}(z+\delta z)\,\delta x\,\delta y - \tau_{xz}(z)\,\delta x\,\delta y\end{aligned}$$

(the water density ρ is taken as constant). Friction forces within the ocean arise especially from eddy shear stress due to eddy fluxes of momentum going through the frictional surface considered (viscous shear stress, however, is generally negligible for large-scale motions), so that according to Table 3.3 $\tau_{xy} = -\rho\,\overline{u'v'}$, $\tau_{xz} = -\rho\,\overline{u'w'}$, except at the boundaries. We obtain for the components of the acceleration,

$$\begin{aligned}a_x &= -\frac{1}{\rho}\frac{\partial p}{\partial x} + \frac{1}{\rho}\frac{\partial \tau_{xy}}{\partial y} + \frac{1}{\rho}\frac{\partial \tau_{xz}}{\partial z} \\ a_y &= -\frac{1}{\rho}\frac{\partial p}{\partial y} + \frac{1}{\rho}\frac{\partial \tau_{yx}}{\partial x} + \frac{1}{\rho}\frac{\partial \tau_{yz}}{\partial z} \\ a_z &= -\frac{1}{\rho}\frac{\partial p}{\partial z} + \frac{1}{\rho}\frac{\partial \tau_{zx}}{\partial x} + \frac{1}{\rho}\frac{\partial \tau_{zy}}{\partial y} - g,\end{aligned} \qquad (6.10)$$

where g denotes the free-fall acceleration, i.e., the resultant acceleration due to the gravity force and the vertical components of centrifugal force and Coriolis force. As the horizontal shears of large-scale ocean circulations are commonly negligibly small compared to the vertical shear we write approximately

$$a_x = -\frac{1}{\rho}\frac{\partial p}{\partial x} + \frac{1}{\rho}\frac{\partial \tau_{xz}}{\partial z} \qquad (6.11a)$$

$$a_y = -\frac{1}{\rho}\frac{\partial p}{\partial y} + \frac{1}{\rho}\frac{\partial \tau_{yz}}{\partial z} \qquad (6.11b)$$

$$a_z = -\frac{1}{\rho}\frac{\partial p}{\partial z} - g. \qquad (6.11c)$$

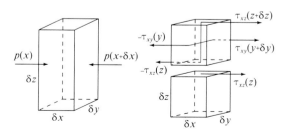

Fig. 6.2 Denominations of the pressure and of the shear stress for the derivation of the pressure gradient accelerations and friction accelerations.

With this we obtain from (6.5) the equation of motion approximated for large-scale horizontal circulations,

$$\frac{Du}{Dt} = -\frac{1}{\rho}\frac{\partial p}{\partial x} + \frac{1}{\rho}\frac{\partial \tau_{xz}}{\partial z} + fv \qquad (6.12a)$$

$$\frac{Dv}{Dt} = -\frac{1}{\rho}\frac{\partial p}{\partial y} + \frac{1}{\rho}\frac{\partial \tau_{yz}}{\partial z} - fu \qquad (6.12b)$$

$$\frac{Dw}{Dt} = -\frac{1}{\rho}\frac{\partial p}{\partial z} - g, \qquad (6.12c)$$

i.e., written out in full with regard to (6.3),

$$\frac{\partial u}{\partial t} + u\frac{\partial u}{\partial x} + v\frac{\partial u}{\partial y} + w\frac{\partial u}{\partial z} = -\frac{1}{\rho}\frac{\partial p}{\partial x} + \frac{1}{\rho}\frac{\partial \tau_{xz}}{\partial z} + fv \qquad (6.13a)$$

$$\frac{\partial v}{\partial t} + u\frac{\partial v}{\partial x} + v\frac{\partial v}{\partial y} + w\frac{\partial v}{\partial z} = -\frac{1}{\rho}\frac{\partial p}{\partial y} + \frac{1}{\rho}\frac{\partial \tau_{yz}}{\partial z} - fu \qquad (6.13b)$$

$$\frac{\partial w}{\partial t} + u\frac{\partial w}{\partial x} + v\frac{\partial w}{\partial y} + w\frac{\partial w}{\partial z} = -\frac{1}{\rho}\frac{\partial p}{\partial z} - g. \qquad (6.13c)$$

6.3 Continuity Equation

The equation system (6.13) is not yet complete. In order to account for the mass conservation, we assume that the ocean water is incompressible and satisfies therefore the continuity equation (3.8):

$$\frac{\partial u}{\partial x} + \frac{\partial v}{\partial y} + \frac{\partial w}{\partial z} = 0. \qquad (6.14)$$

6.4 Special Case: Shallow Water Equations

We now assume that the ocean is a homogeneous layer of water of average thickness H, its surface everywhere at height $z = 0$ in the stationary equilibrium but generally at height $z = \eta$, the ocean bottom at height $z = -H + \eta_b$ (Fig. 6.3). Thus, the local instantaneous layer thickness is $h = H + \eta - \eta_b$. If the average thickness H of the water layer is much smaller than its horizontal extent – a precondition of the subsequently described *shallow water model* – then the vertical accelerations in the water mass will be rather small ($Dw/Dt \approx 0$), so that the ocean will be approximately in the so-called *hydrostatic equilibrium*, defined by the *hydrostatic equation* following from the vertical component of the equation of motion, (6.12c):

$$\frac{\partial p}{\partial z} = -\rho g. \qquad (6.15)$$

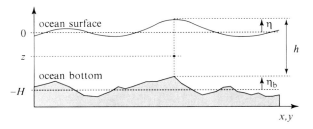

Fig. 6.3 Vertical cross section showing a part of the ocean. Average layer thickness H, water surface at height $z = \eta$, bottom at height $z = -H + \eta_b$, local layer thickness $h = H + \eta - \eta_b$.

Under these conditions, the *hydrostatic pressure* at height z within the water layer is given by

$$p(z) = p(\eta) + \big(p(z) - p(\eta)\big)$$
$$= p(\eta) + \int_\eta^z \frac{\partial p}{\partial z} dz$$
$$= p(\eta) + \rho g (\eta - z).$$

It is therefore equal to the sum of the atmospheric air pressure at the surface of the ocean water, $p(\eta)$, and the weight per unit area of the water column above, $\rho g (\eta - z)$. Assuming the atmospheric air pressure to be constant, we get for the pressure gradients $\partial p/\partial x$ and $\partial p/\partial y$

$$\frac{\partial p}{\partial x} = \rho g \frac{\partial \eta}{\partial x}$$
$$\frac{\partial p}{\partial y} = \rho g \frac{\partial \eta}{\partial y}.$$

Obviously, the pressure gradients within the layer in hydrostatic equilibrium are independent of z. Due to this important fact, the horizontal components of the velocity u and v are constant with height for all time, if this had already been the case at the beginning. Provided this case, the friction forces due to shear stress vanish and the horizontal components of the equation of motion (6.12) become

$$\frac{\partial u}{\partial t} + u \frac{\partial u}{\partial x} + v \frac{\partial u}{\partial y} = -g \frac{\partial \eta}{\partial x} + f v \qquad (6.16a)$$

$$\frac{\partial v}{\partial t} + u \frac{\partial v}{\partial x} + v \frac{\partial v}{\partial y} = -g \frac{\partial \eta}{\partial y} - f u, \qquad (6.16b)$$

where we have written out in full the material derivative, as in (6.13), and the vertical motions neglected.

6.4 Special Case: Shallow Water Equations

In order to close the equation system for the three unknowns u, v and η, we integrate the continuity equation (6.14) from the bottom of the ocean water layer to its surface. To be sufficiently exact, we take the vertical motions into account:

$$\int_{-H+\eta_b}^{\eta} \left(\frac{\partial u}{\partial x} + \frac{\partial v}{\partial y} + \frac{\partial w}{\partial z} \right) dz = \int_{-H+\eta_b}^{\eta} \frac{\partial u}{\partial x} dz + \int_{-H+\eta_b}^{\eta} \frac{\partial v}{\partial y} dz + \int_{-H+\eta_b}^{\eta} \frac{\partial w}{\partial z} dz = 0. \tag{6.17}$$

The limits of integration $z_1 = -H + \eta_b$ and $z_2 = \eta = z_1 + h$ are functions of x and y, i.e. $z_1 = z_1(x, y) = -H + \eta_b(x, y)$ and $z_2 = \eta(t, x, y) = z_1(x, y) + h(t, x, y)$. Using the Leibniz integral rule (rule for the differentiation of a definite integral) and noting that the velocities u and v are constant, we get

$$\int_{-H+\eta_b}^{\eta} \frac{\partial u}{\partial x} dz = \frac{\partial}{\partial x} \int_{-H+\eta_b}^{\eta} u \, dz - u \frac{\partial \eta}{\partial x} + u \frac{\partial \eta_b}{\partial x}$$

$$= \frac{\partial (u(H + \eta - \eta_b))}{\partial x} - u \frac{\partial \eta}{\partial x} + u \frac{\partial \eta_b}{\partial x}$$

$$= \frac{\partial (u h)}{\partial x} - u \frac{\partial \eta}{\partial x} + u \frac{\partial \eta_b}{\partial x}$$

$$\int_{-H+\eta_b}^{\eta} \frac{\partial v}{\partial y} dz = \frac{\partial}{\partial y} \int_{-H+\eta_b}^{\eta} v \, dz - v \frac{\partial \eta}{\partial y} + v \frac{\partial \eta_b}{\partial y}$$

$$= \frac{\partial (v h)}{\partial y} - v \frac{\partial \eta}{\partial y} + v \frac{\partial \eta_b}{\partial y}$$

$$\int_{-H+\eta_b}^{\eta} \frac{\partial w}{\partial z} dz = w(\eta) - w(-H + \eta_b)$$

and with this from (6.17)

$$\frac{\partial (u h)}{\partial x} - u \frac{\partial \eta}{\partial x} + u \frac{\partial \eta_b}{\partial x}$$
$$+ \frac{\partial (v h)}{\partial y} - v \frac{\partial \eta}{\partial y} + v \frac{\partial \eta_b}{\partial y}$$
$$+ w(\eta) - w(-H + \eta_b) = 0. \tag{6.18}$$

The difference between the vertical velocity at the surface $w(\eta)$ and the vertical velocity at the bottom $w(-H + \eta_b)$ corresponds to the change in height per unit time of the water column between $-H + \eta_b$ and η and therefore to the material derivative (with respect to the horizontal motion) of the column height $h(t, x, y) = H + \eta(t, x, y) - \eta_b(x, y)$:

$$w(\eta) - w(-H + \eta_b) = \frac{Dh}{Dt}$$
$$= \frac{\partial h}{\partial t} + u\frac{\partial h}{\partial x} + v\frac{\partial h}{\partial y}$$
$$= \frac{\partial \eta}{\partial t} + u\frac{\partial \eta}{\partial x} + v\frac{\partial \eta}{\partial y} - u\frac{\partial \eta_b}{\partial x} - v\frac{\partial \eta_b}{\partial y}.$$

So, we obtain from (6.18)

$$\frac{\partial \eta}{\partial t} + \frac{\partial (uh)}{\partial x} + \frac{\partial (vh)}{\partial y} = 0$$

or, because $h = H + \eta - \eta_b$, where H as well as η_b are time independent,

$$\frac{\partial h}{\partial t} + \frac{\partial (uh)}{\partial x} + \frac{\partial (vh)}{\partial y} = 0. \tag{6.19}$$

The equation of motion (6.16) and continuity equation (6.19) represent the fundamental equations of the shallow water model, namely the shallow water equations

$$\frac{\partial u}{\partial t} + u\frac{\partial u}{\partial x} + v\frac{\partial u}{\partial y} = -g\frac{\partial \eta}{\partial x} + fv \tag{6.20a}$$

$$\frac{\partial v}{\partial t} + u\frac{\partial v}{\partial x} + v\frac{\partial v}{\partial y} = -g\frac{\partial \eta}{\partial y} - fu \tag{6.20b}$$

$$\frac{\partial h}{\partial t} + u\frac{\partial h}{\partial x} + v\frac{\partial h}{\partial y} = -h\left(\frac{\partial u}{\partial x} + \frac{\partial v}{\partial y}\right), \quad h = H + \eta - \eta_b, \tag{6.20c}$$

and the continuity equation (6.20c) follows from (6.19) with the aid of the product rule of differentiation.

This equation system becomes particularly simple for the case of a non-rotating Earth ($f = 0$), a flat bottom ($\eta_b = 0$), small velocities u, v and elevations $\eta \ll H$ as well as small space derivatives of u, v and h. In this case the Coriolis terms vanish and the non-linear terms are negligible:

$$\frac{\partial u}{\partial t} = -g\frac{\partial \eta}{\partial x} \tag{6.21a}$$

$$\frac{\partial v}{\partial t} = -g\frac{\partial \eta}{\partial y} \tag{6.21b}$$

$$\frac{\partial \eta}{\partial t} = -H\left(\frac{\partial u}{\partial x} + \frac{\partial v}{\partial y}\right). \tag{6.21c}$$

Taking the time derivative of the continuity equation (6.21c), the space derivative $\partial/\partial x$ of the equation of motion (6.21a) and the space derivative $\partial/\partial y$ of the equation of motion (6.21b) we obtain a single equation for the surface elevation:

$$\frac{\partial^2 \eta}{\partial t^2} = g H \vec{\nabla}^2 \eta. \quad (6.22)$$

This is a classical wave equation, formally identical to (3.17). Solutions of the simplified shallow water equations (6.21) are therefore, among others, harmonic dispersion-free waves with phase speed $\sqrt{gH}$. Accounting for the effects of a rotating Earth ($f \neq 0$), i.e. starting from (6.20) and neglecting nonlinear terms, wave equations can be derived which describe Kelvin, Rossby, and planetary-gravity waves.

In order to compute atmospheric and oceanic flows in climate models, spherical coordinates are applied. Hence, the Laplace operator in (6.22) has to be written in spherical coordinates. In the ocean, conditions have to be formulated at the boundaries of ocean basins, in the atmosphere, periodic boundary conditions are postulated.

6.5 Different Types of Grids in Climate Models

The partial differential equations describing the dynamics in climate models need to be discretized. Up to now, we have assumed that all quantities are evaluated at the same grid points. However, in most cases this is not the best choice. It will be shown in simple examples that other arrangements of grids, which represent the physical reality better, lead to much more efficient schemes. This will be illustrated using the one-dimensional version of the simplified shallow water equations (6.21).

The simplified shallow water equations (6.21) in one dimension are given by

$$\frac{\partial u}{\partial t} = -g \frac{\partial \eta}{\partial x} \quad (6.23a)$$

$$\frac{\partial \eta}{\partial t} = -H \frac{\partial u}{\partial x}, \quad (6.23b)$$

where the two unknown functions $u(x,t)$ and $\eta(x,t)$ are to be determined. It is important to realize that the two equations in (6.23) are tightly coupled. If we choose the common discretization in space according to $x = i \Delta x$ with the denominations $u_i \equiv u(i \Delta x, t)$, and $\eta_i \equiv \eta(i \Delta x, t)$, both functions are evaluated at identical grid points (Fig. 6.4). The discretized forms of (6.23) read

$$\frac{\partial u_i}{\partial t} = -g \frac{\eta_{i+1} - \eta_{i-1}}{2 \Delta x}, \quad (6.24a)$$

$$\frac{\partial \eta_i}{\partial t} = -H \frac{u_{i+1} - u_{i-1}}{2 \Delta x}. \quad (6.24b)$$

Thus, it appears that the two schemes are applied on two independent sub-grids, the solution vectors (η_{2k}, u_{2k+1}) and (η_{2k+1}, u_{2k}) are mutually independent and no information is interchanged. The error of the schemes in (6.24) is of order Δx^2.

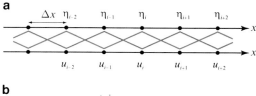

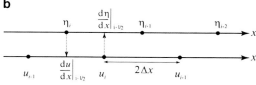

Fig. 6.4 (a) Simple grid for the shallow water equations. All functions are evaluated at the same points. Two independent sub-grids, connected with the red and blue lines, result. (b) Staggered grid for the shallow water equations. Flux quantities (u) and volume quantities (η) are evaluated at different points.

By shifting one axis in Fig. 6.4a, we consider a *staggered grid* as it is shown in Fig. 6.4b. Here, twice the grid spacing as before is chosen. Therefore, only half the number of values needs to be computed. The discretized forms of (6.23) for this grid are given by

$$\frac{\partial u_i}{\partial t} = -g \left.\frac{\partial \eta}{\partial x}\right|_{i+1/2} = -g \frac{\eta_{i+1} - \eta_i}{2\,\Delta x}, \qquad (6.25\text{a})$$

$$\frac{\partial \eta_i}{\partial t} = -H \left.\frac{\partial u}{\partial x}\right|_{i-1/2} = -H \frac{u_i - u_{i-1}}{2\,\Delta x}. \qquad (6.25\text{b})$$

By evaluating derivatives in (6.25) at the intermediate points, they can be regarded as central differences with an equivalent grid spacing of Δx, even though indices only include immediate neighbors. For this reason, the schemes in (6.25) are of the same accuracy as the ones in (6.24), where twice the number of values need to be computed. Hence, the staggered grid affords a significant improvement with regard to the present differential equations.

These findings can be generalized to two dimensions. To illustrate this, we consider again the equation system (6.21), where the three unknown functions $u(x, y, t)$, $v(x, y, t)$ and $\eta(x, y, t)$ have to be computed on a two-dimensional grid (Fig. 6.5). In case all functions are evaluated at the same grid points (Fig. 6.5a), we denote this an A-grid (*Arakawa A-grid*). An alternative choice is an E-grid (*Arakawa E-grid*) for which the velocity components are evaluated at points between the η points (Fig. 6.5b) and in so doing the relations between the velocity components and the horizontal gradients of η are taken into account.

A further commonly used grid is the C-grid (*Arakawa C-grid*), for which the velocity components of different directions are evaluated at different grid points (Fig. 6.6). This grid structure represents the physics of fluid motion most

6.5 Different Types of Grids in Climate Models

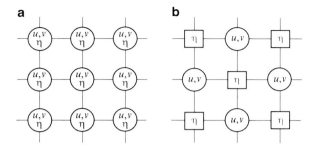

Fig. 6.5 (a) Two-dimensional Arakawa A-grid, in which all functions are evaluated at identical grid points. (b) Two-dimensional Arakawa E-grid, where flux quantities (u, v) and volume quantities (η) are evaluated at different places.

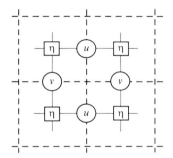

Fig. 6.6 Two-dimensional Arakawa C-grid, in which flux quantities of different directions are evaluated at different points, while volume quantities are computed in between. The *dashed grid* illustrates the physical meaning of the C-grid. Volume quantities (η) are located in the centers of the *dashed boxes*, while flux quantities (u, v) are centered on the box boundaries. Therefore, the C-grid accounts for mass balance in a natural way.

appropriately, because *flux quantities* (e.g., velocities, energy fluxes, etc.) are defined at the boundary of grid boxes while *volume quantities* (e.g., surface elevation, concentration, temperature, etc.) are represented in the center.

The question concerning the grid type also plays a role in the solution of the one-dimensional energy balance model. Equation (4.9) can be simplified to

$$\frac{\partial T}{\partial t} = a + b\,T^4 + c\,\frac{\partial}{\partial \varphi}\left(e\,\frac{\partial T}{\partial \varphi}\right) \tag{6.26}$$

with spatially dependent coefficients a, b, c and e. In this model, temperature is the volume quantity while the meridional temperature gradient represents a flux quantity. If an A-grid is selected (Fig. 6.7a), the discretized form of (6.26) reads

$$\frac{\partial T_i}{\partial t} = a_i + b_i\,T_i^4 + c_i\,\frac{e_{i+1}\,T'_{i+1} - e_{i-1}\,T'_{i-1}}{2\,\Delta \varphi} \tag{6.27a}$$

$$T'_i = \frac{T_{i+1} - T_{i-1}}{2\,\Delta \varphi}, \tag{6.27b}$$

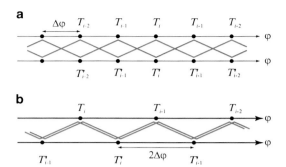

Fig. 6.7 (a) A-grid for the one-dimensional energy balance model. The scheme (6.27) for the solution of (6.26) results in two independent sub-grids (*colored lines*). (b) C-grid for the one-dimensional energy balance model.

where again (T_{2k}, T'_{2k+1}) and (T_{2k+1}, T'_{2k}) are independent solution vectors. The solution is evaluated on two non-connected sub-grids.

In a C-grid configuration (Fig. 6.7b) with double grid spacing, only half the number of functions has to be evaluated and, accordingly, the discretized form reads

$$\frac{\partial T_i}{\partial t} = a_i + b_i T_i^4 + c_i \frac{e_i T'_i - e_{i-1} T'_{i-1}}{2 \Delta\varphi} \tag{6.28a}$$

$$T'_i = \frac{T_{i+1} - T_i}{2 \Delta\varphi} \tag{6.28b}$$

which is of the same accuracy, but requires only half the computational resources. In addition, the implementation of boundary conditions with respect to the flux quantities (see (4.10)) is straightforward, since they can be set to zero: $T'_0 = 0$ and $T'_M = 0$.

6.6 Spectral Models

Here, a short section on an important alternative method to solve partial differential equations in spherical geometry is presented. Up to now, we have treated several methods that make use of finite differences. For global climate models, the integration domain covers a sphere, which enables the use of particular functions for the solution of the partial differential equations. Therefore, in order to solve equations of the type given in (6.22) on a sphere, spectral methods are often applied.

Usually, the atmospheric components of global climate models are spectral models. In global ocean models they are employed rarely, or only for the vertical component as the strong gradients of properties near the surface (e.g., temperature) can be better accounted for.

6.6 Spectral Models

Instead of spanning a grid over the sphere and then replacing the differential equations by a system of equations in finite differences, the unknown functions are expanded by appropriate basis functions which satisfy certain boundary conditions.

Consider eigenfunctions of the Laplace operator on a sphere of radius R,

$$\vec{\nabla}^2 Y_\ell^m = -\frac{\ell(\ell+1)}{R^2} Y_\ell^m, \tag{6.29}$$

namely *spherical harmonics*, which are given by

$$Y_\ell^m(\varphi, \lambda) = P_\ell^m(\sin\varphi) e^{im\lambda}, \tag{6.30}$$

where $P_\ell^m(\sin\varphi)$ are *associated Legendre functions of the 1st kind*. The quantities m and ℓ are wave numbers: $2m$ is the number of knot meridians (zeroes on a circle of latitude), $\ell - m$ is the number of knot latitudes excluding the two poles. The following orthogonality relation is valid

$$\frac{1}{4\pi} \int_{-1}^{1} d(\sin\varphi) \int_{0}^{2\pi} d\lambda\, Y_\ell^m Y_{\ell'}^{m'} = \begin{cases} 1 & \text{if } m = m', \ell = \ell' \\ 0 & \text{else} \end{cases} \tag{6.31}$$

which is consistent with the fact, that (6.30) constitutes a complete basis of functions.

The unknown solution of (6.22) is now expressed as a linear combination of basis functions $Y_\ell^m(\varphi, \lambda)$ with time-dependent coefficients $\Phi_\ell^m(t)$:

$$\eta(t, \varphi, \lambda) = \sum_{|m|\leq\ell} \sum_{\ell} \Phi_\ell^m(t) Y_\ell^m(\varphi, \lambda). \tag{6.32}$$

Inserting (6.32) into (6.22) and using (6.29) we obtain following ordinary differential equations for the coefficient functions:

$$\frac{d^2 \Phi_\ell^m}{dt^2} = -\ell(\ell+1) \frac{gH}{R^2} \Phi_\ell^m. \tag{6.33}$$

Hence, the partial differential equation (6.22) is replaced by a set of ordinary differential equations for the coefficient functions $\Phi_\ell^m(t)$.

The expansion in (6.32) theoretically ranges from $\ell = 0, \ldots, \infty$, $m = -\ell, \ldots, +\ell$, but in practice, the summation needs to be truncated at an appropriate point. This results in finite spatial resolution determined by the highest wavenumbers. The most commonly used truncations are *triangular* and *rhomboidal* truncations, schematically illustrated in Fig. 6.8.

Early GCMs used R15 and R21. Transient eddies, important features of the atmosphere, are barely resolved in R15. Hence, the partitioning – in absolute terms – of

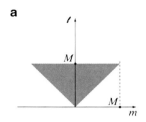

$$\eta(t,\varphi,\lambda) = \sum_{m=-M}^{M} \sum_{\ell=|m|}^{M} \Phi_\ell^m(t) Y_\ell^m(\varphi,\lambda)$$

denomination: T(M)

examples: T21, T31, T42, T63, T85, T106.

Fig. 6.8a Triangular truncation.

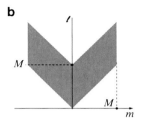

$$\eta(t,\varphi,\lambda) = \sum_{m=-M}^{M} \sum_{\ell=|m|}^{|m|+M} \Phi_\ell^m(t) Y_\ell^m(\varphi,\lambda)$$

denomination: R(M)

examples: R15, R21, R30.

Fig. 6.8b Rhomboidal truncation.

the meridional heat transport in the atmosphere is not realistically simulated. This is one of the reasons for coupled models of low resolution to require flux corrections (see also Sect. 8.6). Simulations are currently performed at T42 to T85. Results of a simulation with very high resolution (T106) were shown in Fig. 2.3.

The choice of the basis function already satisfies some of the boundary conditions. This is a distinct advantage of spectral models. However, one difficulty arises with the treatment of the non-linear terms and terms describing Coriolis effects which are part of the full equations of motion. When these effects are considered, spectral models become much more complicated, and coupling between the individual wave numbers occurs.

6.7 Wind-Driven Flow in the Ocean (Stommel Model)

Since the beginning of inter-continental marine navigation in the fifteenth century, it is well known that the surface flow in the ocean is characterized by large-scale gyres (in the northern hemisphere clock-wise subtropical gyre, counter-clockwise subpolar gyre). These gyres are not spatially uniform but feature a strongly intensified current along the western boundary of the ocean basin, namely a strong northward current in the northern hemisphere and a strong southward current in the southern hemisphere, while in the eastern part the currents are weak.

The well-known Gulf stream is part of the western part of the North Atlantic's subtropical gyre. This then turns into the Transatlantic Drift Current as soon as it leaves the American East Coast and moves northward towards the eastern part of the subpolar gyre. Its effects on temperature and salinity are observed as far as north Spitsbergen. The Kuroshio Current, the Brazil Current and others form dynamically similar circulation systems.

6.7 Wind-Driven Flow in the Ocean (Stommel Model)

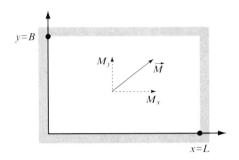

Fig. 6.9 Geometry of the ocean basin and mass transport in a Cartesian coordinate system, which is tangent to the Earth's sphere at a given latitude.

The wind, i.e., the Westerlies in the mid-latitudes and the pronounced Easterlies more towards the equator, have quickly been identified as causes of these currents. However, the dynamical problem, why the ocean currents only intensify along the western basin boundaries, has not been resolved until 1948 when a landmark paper was published by Henry Stommel (1948). Using an elegant model, he demonstrated that the spherical shape of the rotating Earth is the origin of this conspicuous phenomenon.

Following Stommel we consider a homogenous fluid (ρ = constant) in a flat rectangular basin (Fig. 6.9) on the β-plane; vertical cross-section as shown in Fig. 6.3. We assume the bottom to be flat, $\eta_b = 0$, further the atmospheric air pressure at the surface of the ocean water $p(\eta)$ to be constant, and finally the vertical elevation to be much smaller than the mean layer thickness, i.e. $\eta \ll H$. Multiplication of the horizontal components of the equation of motion (6.13) with ρ, integration over the entire depth from the height of the bottom $z = -H$ to the height of the water surface $z = \eta$ plus the assumption of stationarity $\partial/\partial t = 0$ and linearity yields

$$-f \int_{-H}^{\eta} \rho v \, dz = -\int_{-H}^{\eta} \frac{\partial p}{\partial x} \, dz + \tau_{xz}(\eta) - \tau_{xz}(-H), \qquad (6.34a)$$

$$f \int_{-H}^{\eta} \rho u \, dz = -\int_{-H}^{\eta} \frac{\partial p}{\partial y} \, dz + \tau_{yz}(\eta) - \tau_{yz}(-H). \qquad (6.34b)$$

We define the mass transport as follows:

$$\vec{M} = \int_{-H}^{\eta} \rho \vec{u} \, dz \qquad (6.35)$$

and substitute this in (6.34). Equation (6.34) reveals that the mass transport is driven by the shear at the surface and slowed by the friction on the ground. Hence, at the surface the effect of the wind is to transfer momentum into the fluid. The flux of momentum must be passed on to the fluid by internal friction or friction at the bottom of the ocean basin.

Stommel chose the simplest possible parameterisation for this effect by postulating that the shear exerted by the bottom is proportional to the velocity, or the mass transport, respectively. Hence, (6.34) becomes

$$-f M_y = -\int_{-H}^{\eta} \frac{\partial p}{\partial x}\, dz + \tau_{xz}(\eta) - R M_x, \qquad (6.36a)$$

$$f M_x = -\int_{-H}^{\eta} \frac{\partial p}{\partial y}\, dz + \tau_{yz}(\eta) - R M_y, \qquad (6.36b)$$

where R is an inverse characteristic time during which the current comes to rest due to friction.

6.7.1 Determination of the Stream Function

By cross-differentiation $\partial(6.36b)/\partial x - \partial(6.36a)/\partial y$ the pressure gradient terms in (6.36) are eliminated approximately due to $\eta \ll H$. Taking (6.9) into account, we obtain

$$\beta M_y + f \left(\frac{\partial M_x}{\partial x} + \frac{\partial M_y}{\partial y} \right) = \frac{\partial \tau_{yz}}{\partial x} - \frac{\partial \tau_{xz}}{\partial y} - R \left(\frac{\partial M_y}{\partial x} - \frac{\partial M_x}{\partial y} \right), \qquad (6.37)$$

where the functions τ_{xz} and τ_{yz} are now written without argument.

The two unknown components of the mass transport are not mutually independent, since in a closed basin mass conservation must be satisfied. The vertical integration of continuity equation (6.14) yields, analogously to the derivation leading to (6.18) but with $\partial/\partial t = 0$,

$$\vec{\nabla} \cdot \vec{M} = \frac{\partial M_x}{\partial x} + \frac{\partial M_y}{\partial y} = 0, \qquad (6.38)$$

where the unknown vector function $\vec{M}$ can now be replaced by a scalar choosing

$$M_x = -\frac{\partial \Psi}{\partial y}, \qquad (6.39a)$$

$$M_y = \frac{\partial \Psi}{\partial x}. \qquad (6.39b)$$

The scalar function $\Psi(x, y)$ is called *stream function*. Streamlines are lines of constant stream function, along which the current moves tangentially.

Definition (6.39) satisfies (6.38) automatically, and we can use (6.39) in (6.37) in order to obtain the Stommel equation which was first formulated in 1948

6.7 Wind-Driven Flow in the Ocean (Stommel Model)

(Stommel 1948):

$$\beta \frac{\partial \Psi}{\partial x} = \frac{\partial \tau_{yz}}{\partial x} - \frac{\partial \tau_{xz}}{\partial y} - R \left(\frac{\partial^2 \Psi}{\partial x^2} + \frac{\partial^2 \Psi}{\partial y^2} \right). \tag{6.40}$$

This equation contains the phenomenon of western boundary currents in an ocean basin in principle. Equation (6.40) is a partial differential equation of 2nd order in x and y for the function $\Psi(x, y)$.

Boundary conditions still remain to be formulated. Since the transport must be parallel to the boundaries, we require along the boundaries in the y-direction $M_x = -\partial \Psi / \partial y = 0$, and along the boundaries in the x-direction $M_y = \partial \Psi / \partial x = 0$. Hence, Ψ is constant along the boundary. Because (6.40) only contains derivatives of Ψ, we can set, without loss of generality,

$$\Psi = 0 \quad \text{at the boundaries.} \tag{6.41}$$

Therefore, the Stommel model is a boundary value problem with *Dirichlet boundary conditions* (Sect. 5.1). In order to find the solution, the wind stress must be prescribed. For particularly simple spatial relationships of the stress, the boundary problem may even be solved analytically. To this end, Stommel chose a purely zonal wind stress given by

$$\tau_{xz} = -T \cos\left(\frac{\pi}{B} y \right), \tag{6.42a}$$

$$\tau_{yz} = 0. \tag{6.42b}$$

Thus, (6.40) can be solved analytically by separation of the variables. But for more complicated profiles of the wind stress, numerical methods, presented in Chap. 5, need to be applied. We will not explain this analytical solution but are going to discuss numerical solutions of this problem.

The numerical solutions of the boundary value problem (6.40), (6.41) in a rectangular basin between $0 \leq x \leq 7{,}000$ km and $0 \leq y \leq 5{,}000$ km are illustrated in Fig. 6.10. We have employed the method of successive overrelaxation described in Sect. 5.3.2. On a β-plane, a western boundary current develops; for $\beta = 0$, a symmetric solution results which exhibits no boundary current. The western boundary current in this model appears as soon as the Coriolis parameter f depends on the latitude, implying that the spherical shape of the Earth plays a fundamental role in the establishment of the dynamics.

In case a boundary current is present, the x derivatives of the stream function in (6.40) become dominant at the boundary. Assuming a typical lateral width δ of the boundary current and inserting $\Psi \sim 1 - e^{-x/\delta}$ into (6.40), we obtain

$$\beta \frac{1}{\delta} \sim R \frac{1}{\delta^2} \quad \text{hence} \quad \delta \sim \frac{R}{\beta}. \tag{6.43}$$

The width of the boundary current (Stommel boundary layer) scales with the friction coefficient and is inversely proportional to β.

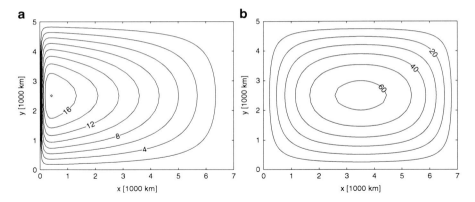

Fig. 6.10 Stream function Ψ (in Sverdrup, $1\,\text{Sv} = 10^6\,\text{m}^3\,\text{s}^{-1}$) of the Stommel model (6.40) for (**a**) $\beta = 2 \cdot 10^{-11}\,\text{m}^{-1}\,\text{s}^{-1}$ and (**b**) $\beta = 0$, with $R = 1/(6\,\text{days})$, and $T = 0.1\,\text{N}\,\text{m}^{-2}$ in (6.42). The numerical solution was computed on a grid with $(N_x = 100) \times (N_y = 20)$ and using the method of successive overrelaxation (5.22). The current flows clockwise and is parallel to the stream lines.

6.7.2 Determination of the Water Surface Elevation

According to (6.36), wind-driven flow induces pressure gradients, which become manifest as an elevation η of the water surface. This effect shall be quantified in the following. It will lead to a boundary value problem with *Neumann boundary conditions* (Sect. 5.1).

Analogously to (6.35), we define the pressure integrated over the depth as

$$P = \int_{-H}^{\eta} p \, dz \qquad (6.44)$$

and take $\partial(6.36a)/\partial x + \partial(6.36b)/\partial y$ with regard to $\eta \ll H$. Using (6.38) and (6.39), for $P(x, y)$ we obtain now the following Poisson equation

$$\vec{\nabla}^2 P = f\,\vec{\nabla}^2 \Psi + \beta\,\frac{\partial \Psi}{\partial y} + \frac{\partial \tau_{xz}}{\partial x} + \frac{\partial \tau_{yz}}{\partial y}. \qquad (6.45)$$

The previous choice of the wind stress (6.42) allows us to cancel the last two terms in (6.45). The boundary conditions for $P(x, y)$ may be derived from (6.36) and the fact that the transport must be parallel to the boundaries:

$$\frac{\partial P}{\partial x} = f\,\frac{\partial \Psi}{\partial x} + \tau_{xz} \qquad \text{at} \quad x = 0 \quad \text{and} \quad x = L \qquad (6.46a)$$

$$\frac{\partial P}{\partial y} = f\,\frac{\partial \Psi}{\partial y} + \tau_{yz} \qquad \text{at} \quad y = 0 \quad \text{and} \quad y = B. \qquad (6.46b)$$

6.7 Wind-Driven Flow in the Ocean (Stommel Model)

Consequently, the derivatives of P perpendicular to the boundary are fixed (Neumann boundary conditions). It must be noted, that (6.45) and (6.46) restrict the solution up to a single constant.

By calculating $P(x, y)$ based on (6.45) and considering (6.46), we can determine the elevation of the water surface using (6.44) and assuming hydrostatic equilibrium:

$$P(x, y) = \int_{-H}^{\eta} \rho g \, (\eta - z) \, dz = \frac{1}{2} \rho g \, (H + \eta)^2 . \tag{6.47}$$

We expand (6.47) with regard to $\eta \ll H$,

$$P(x, y) = \frac{1}{2} \rho g \, H^2 \left(1 + \frac{\eta}{H}\right)^2$$
$$\approx \frac{1}{2} \rho g \, H^2 + \rho g \, H \, \eta,$$

and find with this

$$\eta(x, y) \approx \frac{P(x, y)}{\rho g \, H} - \frac{H}{2}. \tag{6.48}$$

The numerical solution of a boundary value problem with Neumann boundary conditions requires some additional considerations. For Dirichlet boundary conditions, such as (6.41), the boundary values are accounted for naturally by setting the values in the numerical scheme directly. However, Neumann boundary conditions require additional information from the points next to the boundary in order to find the values at the boundary itself.

We derive the discretized schemes to determine the boundary values in the case of Neumann boundary conditions. The idea is to calculate the derivatives at the boundary using the values of the grid points inside and assuming an appropriate interpolation. There are various possibilities for this: linear, parabolic, etc. We explain the approach for the boundaries $x = 0$ and $x = L$; corresponding formulations for the other boundaries can be inferred analogously.

In x-direction, the discretisation $\Delta x = L/N$, with $x = i \, \Delta x$ is chosen. We evaluate the solution function $P(x)$ at the grid points, that is $P(i \, \Delta x) \equiv P_i$, where P_0 and P_N are located at the respective boundaries ($x = 0$ and $x = L$, Fig. 6.11). A parabola is assumed to interpolate the solution between the boundary point and

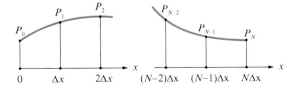

Fig. 6.11 Interpolation of the solution function at the boundary using parabolas.

the two points closest to the boundary. For the boundary $x = 0$, we assume the quadratic function

$$y = a x^2 + b x + P_0. \tag{6.49}$$

In order to assure that the parabola goes through the values P_1 and P_2, the following must be valid

$$P_1 = a (\Delta x)^2 + b (\Delta x) + P_0 \quad \text{and} \quad P_2 = a (2 \Delta x)^2 + b (2 \Delta x) + P_0, \tag{6.50}$$

and analogous expressions hold for the boundary at $x = L$. Solving (6.50) for the coefficients of the interpolation parabola we obtain

$$a = \frac{P_2 - 2 P_1 + P_0}{2 \Delta x^2}, \tag{6.51a}$$

$$b = \frac{-P_2 + 4 P_1 - 3 P_0}{2 \Delta x}. \tag{6.51b}$$

With this, the first derivative at the boundary can be computed using (6.49):

$$\left.\frac{dy}{dx}\right|_{x=0} = b. \tag{6.52}$$

Hence, for the derivative to be given as a boundary condition at the boundaries, we can apply (6.51) and (6.52) in order to calculate the value of the function at the boundary. We find

$$P_0 = \frac{4 P_1 - P_2}{3} - \frac{2}{3} \Delta x \left.\frac{dy}{dx}\right|_{x=0} \tag{6.53a}$$

$$P_N = \frac{4 P_{N-1} - P_{N-2}}{3} + \frac{2}{3} \Delta x \left.\frac{dy}{dx}\right|_{x=L}. \tag{6.53b}$$

The numerical solution of (6.45), shown for different parameter values in Fig. 6.12, was computed inside the domain using the method of successive overrelaxation according to (5.22). Therefore, $\Psi(x, y)$ needs to be determined first by solving the Dirichlet boundary value problem given by (6.40) and (6.41). The boundary conditions (6.46) are accounted for by computing the boundary values according to (6.53).

The current is clock-wise (Fig. 6.10). Inside the western boundary current, pressure gradient, Coriolis and inertial forces are in equilibrium with the wind stress (Fig. 6.12a). On an f-plane ($\beta = 0$, Fig. 6.12b), the current is approximately in a geostrophic equilibrium (Coriolis forces are balanced mainly by the pressure gradients, friction compensates for the wind stress). It must be noted, that due to the friction, currents do not exactly follow the lines of constant pressure, although

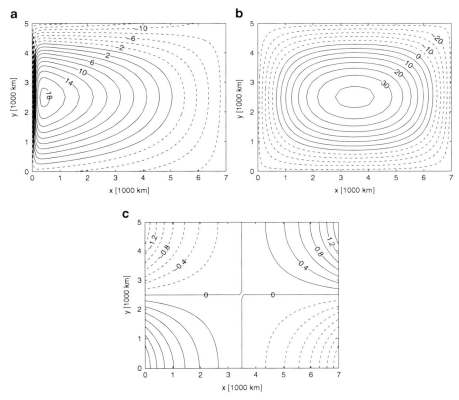

Fig. 6.12 Surface elevation η (in centimeter) calculated using the Stommel model (6.40). Panel (**a**) $\beta = 2 \cdot 10^{-11}\,\mathrm{m^{-1}\,s^{-1}}$; Panel (**b**) $\beta = 0$, and Panel (**c**) $f = 0$. The parameters are $R = 1/(6\,\mathrm{days})$, $H = 1{,}000\,\mathrm{m}$ and $T = 0.1\,\mathrm{N\,m^{-2}}$ in (6.42). The numerical solution of (6.45) and (6.46) was calculated on a grid with $(N_x = 100) \times (N_y = 20)$ using the method of successive overrelaxation (5.22).

$\nabla^2 P = f\,\vec{\nabla}^2 \Psi$ is valid inside the domain. This follows from the equation of motion (6.36). In case the reference system is not rotating ($f = 0$, Fig. 6.12c), the meridional flow is directed parallel to the negative pressure gradients, i.e. "downhill", and the zonal flow is forced to flow "uphill", i.e. against the pressure gradient, owing to the zonal wind stress.

6.8 Potential Vorticity: An Important Conserved Quantity

Conservation theorems are fundamental statements in physics and enable a more profound understanding of various processes responsible for the dynamics. Hence, conservation theorems and related quantities are also very useful in geophysical

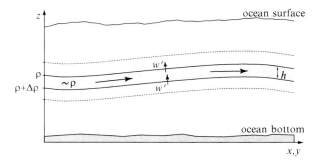

Fig. 6.13 Vertical cross section showing a part of a discontinuously stratified ocean. Providing that the stratification is stable, the water of the lower layers are denser than the water of the upper layers ($\Delta \rho > 0$). Eddy fluctuations w' of the vertical velocity w occur and yield mean eddy mass fluxes $\pm \overline{w' \rho'} = \pm \overline{w' \Delta \rho}$ between the layers.

fluid dynamics and climate modelling. A conservation equation for large-scale ocean flow is derived from the equations of motion (6.13) in this section.

The following explanations of this section are based on a simple model of a large-scale ocean flow in hydrostatic equilibrium (Sect. 6.4). It approximates the continuous stratification of the real ocean water by a discontinuous stratification formed by superimposed thin layers, shallow water layers indeed, as illustrated in Fig. 6.13. Any of these layers has a constant density ρ and a variable thickness $h(x, y)$ and slides between the underlying denser layer and the overlying lighter layer, thereby moving along surfaces of constant density (isopycnals). The function $\rho h(x, y)$ represents the mass per unit area in this layer and obeys the relation

$$\frac{\partial (\rho h)}{\partial t} + \frac{\partial}{\partial x} (u \rho h) + \frac{\partial}{\partial y} (v \rho h) = Q. \qquad (6.54)$$

This is a generalized version of the continuity equation (6.19) of the shallow water model taking into account a *cross-isopycnal mass flux* Q (in $\mathrm{kg\,m^{-2}\,s^{-1}}$) as well, which could arise for example from eddy mass fluxes $\pm \overline{w' \rho'}$ (covariance between vertical velocity w and density ρ) going through the upper and the lower boundaries of the layer. Using definition (6.3) for horizontal motions and neglecting density changes (but not volume changes) within the layer, (6.54) can be written as

$$\frac{Dh}{Dt} + h \left(\frac{\partial u}{\partial x} + \frac{\partial v}{\partial y} \right) = \frac{Q}{\rho}. \qquad (6.55)$$

We now define the *vorticity* measured relatively to the Earth's surface, namely the *relative vorticity* ζ, as the vertical component of the curl of the velocity field $\vec{u}$, which is measured relatively to the Earth's surface, according to

$$\zeta = \left(\vec{\nabla} \times \vec{u} \right) \cdot \hat{z} = \frac{\partial v}{\partial x} - \frac{\partial u}{\partial y}. \qquad (6.56)$$

6.8 Potential Vorticity: An Important Conserved Quantity

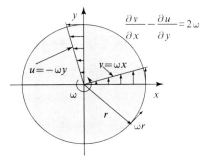

Fig. 6.14 Vortex in the form of a solid disk rotating with angular velocity ω about an Earth-fixed z-axis. The velocity of the points of the vortex at a distance r from the center is ωr and always tangential. Hence, the relative vorticity is 2ω.

$\hat{z}$ is the unit vector normal to the Earth's surface. It can be shown that the relative vorticity ζ equals twice the angular velocity of an infinitesimal vortex on Earth as illustrated in Fig. 6.14.

To examine the time evolution of relative vorticity inside the layer, we consider the equations of motion (6.13) – namely a generalized version of the equations of motion of the shallow water model (6.20) – and calculate $\partial/\partial y$ (6.13a) and $\partial/\partial x$ (6.13b) assuming $w = 0$. Observing (6.11) we obtain

$$\frac{\partial}{\partial y}(6.13a): \frac{\partial}{\partial t}\frac{\partial u}{\partial y} + \frac{\partial u}{\partial y}\frac{\partial u}{\partial x} + u\frac{\partial^2 u}{\partial x \partial y} + \frac{\partial v}{\partial y}\frac{\partial u}{\partial y} + v\frac{\partial^2 u}{\partial y^2} = \frac{\partial a_x}{\partial y} + \frac{\partial f}{\partial y}v + f\frac{\partial v}{\partial y},$$

$$\frac{\partial}{\partial x}(6.13b): \frac{\partial}{\partial t}\frac{\partial v}{\partial x} + \frac{\partial u}{\partial x}\frac{\partial v}{\partial x} + u\frac{\partial^2 v}{\partial x^2} + \frac{\partial v}{\partial x}\frac{\partial v}{\partial y} + v\frac{\partial^2 v}{\partial y \partial x} = \frac{\partial a_y}{\partial x} - \frac{\partial f}{\partial x}u - f\frac{\partial u}{\partial x},$$

so that $\partial/\partial x$ (6.13b) $- \partial/\partial y$ (6.13a) reads

$$\frac{\partial}{\partial t}\left(\frac{\partial v}{\partial x} - \frac{\partial u}{\partial y}\right) + u\frac{\partial}{\partial x}\left(\frac{\partial v}{\partial x} - \frac{\partial u}{\partial y}\right) + v\frac{\partial}{\partial y}\left(\frac{\partial v}{\partial x} - \frac{\partial u}{\partial y}\right)$$
$$+ \frac{\partial u}{\partial x}\left(\frac{\partial v}{\partial x} - \frac{\partial u}{\partial y}\right) + \frac{\partial v}{\partial y}\left(\frac{\partial v}{\partial x} - \frac{\partial u}{\partial y}\right)$$
$$= \frac{\partial a_y}{\partial x} - \frac{\partial a_x}{\partial y} - \frac{\partial f}{\partial x}u - f\frac{\partial u}{\partial x} - \frac{\partial f}{\partial y}v - f\frac{\partial v}{\partial y}$$

and with (6.56) and $\partial f/\partial t = 0$

$$\frac{\partial \zeta}{\partial t} + u\frac{\partial \zeta}{\partial x} + v\frac{\partial \zeta}{\partial y} + \left(\frac{\partial u}{\partial x} + \frac{\partial v}{\partial y}\right)\zeta$$
$$= \frac{\partial a_y}{\partial x} - \frac{\partial a_x}{\partial y} - \frac{\partial f}{\partial t} - u\frac{\partial f}{\partial x} - v\frac{\partial f}{\partial y} - f\left(\frac{\partial u}{\partial x} + \frac{\partial v}{\partial y}\right),$$

i.e.

$$\frac{D}{Dt}(\zeta + f) = \underbrace{-(\zeta + f)\left(\frac{\partial u}{\partial x} + \frac{\partial v}{\partial y}\right)}_{\text{CON}} + \underbrace{\frac{\partial a_y}{\partial x} - \frac{\partial a_x}{\partial y}}_{\text{PRO}}. \tag{6.57}$$

$\zeta + f$ is the *absolute vorticity*, i.e. the vorticity taken relative to an unaccelerated reference system ($f = 2\Omega \sin\varphi$ is the vorticity of the rotating surface of the Earth at latitude φ).

We consider the terms on the right-hand side of (6.57). They signify two distinct sources of absolute vorticity: (a) convergence of the flow (CON), and (b) production by real forces (PRO). From (6.11) it follows for ocean water with constant density ρ:

$$\frac{\partial a_y}{\partial x} - \frac{\partial a_x}{\partial y} = \frac{1}{\rho}\frac{\partial^2 \tau_{yz}}{\partial x \, \partial z} - \frac{1}{\rho}\frac{\partial^2 \tau_{xz}}{\partial y \, \partial z}.$$

In such a fluid the production by real forces (PRO) is independent of the pressure gradient forces. Such a fluid is called *barotropic*, all the others are *baroclinic* (Sect. 4.4). In a barotropic fluid the change of the absolute vorticity results solely from circulation convergence and vorticity production due to friction.

Equation (6.55) allows us to simplify the term CON in (6.57) applied to the shallow water layer emphasized in Fig. 6.13:

$$\frac{D}{Dt}(\zeta + f) = -\frac{\zeta + f}{h}\left(\frac{Q}{\rho} - \frac{Dh}{Dt}\right) + \frac{\partial a_y}{\partial x} - \frac{\partial a_x}{\partial y},$$

i.e.

$$\frac{1}{h}\frac{D}{Dt}(\zeta + f) - \frac{\zeta + f}{h^2}\frac{Dh}{Dt} = -\frac{\zeta + f}{h}\frac{Q}{\rho h} + \frac{1}{h}\left(\frac{\partial a_y}{\partial x} - \frac{\partial a_x}{\partial y}\right)$$

and consequently

$$\frac{D}{Dt}\left(\frac{\zeta + f}{h}\right) = -\frac{\zeta + f}{h}\frac{Q}{\rho h} + \frac{1}{h}\left(\frac{\partial a_y}{\partial x} - \frac{\partial a_x}{\partial y}\right). \tag{6.58}$$

The quantity $(\zeta + f)/h$ is the *potential vorticity* in the shallow water layer. Potential vorticity is a conservative quantity in a barotropic and frictionless ocean circulation, if no mass is supplied or removed.

Regarding (6.58), wind-driven flow described in Sect. 6.7 can now be understood in a coherent framework. In the Stommel model, a closed flat basin ($Q = 0$ and $\eta_b = 0$, Fig. 6.3) with only one shallow water layer was considered. We integrate (6.58) over the layer thickness h, assuming $\eta \ll H$ (so that $h = H + \eta \approx H =$ constant), and substitute a_x and a_y for the right-hand side of (6.36). We assume $\tau_{yz} = 0$ according to the Stommel model. This results in the approximation

6.8 Potential Vorticity: An Important Conserved Quantity

Table 6.1 Signs of the individual terms in (6.59) for the Stommel model in the northern hemisphere. The relation shaded in grey is required in order to close the balance of terms. The large gradients imply a strong, confined flow, i.e., a boundary current.

Direction of flow	1 $\rho h \dfrac{D\zeta}{Dt}$	2 $+\rho h \dfrac{Df}{Dt}$	3 $\approx -\dfrac{\partial \tau_{xz}}{\partial y}$	4 $-R\dfrac{\partial M_y}{\partial x}$	5 $+R\dfrac{\partial M_x}{\partial y}$
N → S	≈ 0	< 0	< 0	≈ 0	≈ 0
S → N	≈ 0	> 0	< 0	≫ 0	≈ 0

$$\rho h \frac{D\zeta}{Dt} + \rho h \frac{Df}{Dt} \approx -\frac{\partial \tau_{xz}}{\partial y} - R\frac{\partial M_y}{\partial x} + R\frac{\partial M_x}{\partial y}. \qquad (6.59)$$

An estimate for the individual terms in (6.59) for large-scale circulation of typical spatial scales of 10^6 m reveals the individual contributions given in Table 6.1 and provides substantial insight into the dynamics of large-scale geophysical flow.

We now consider the signs and magnitudes of the five terms in (6.59) for northward and southward flow. The dominant term on the left-hand side is Df/Dt (term 2 in Table 6.1), and the material derivative of the relative vorticity (term 1) can be neglected in comparison. Southward flow implies decreasing f, and for northward flow f increases. The sign of term 3 is always negative, and the west-east mass transport M_x (term 5) vanishes towards the eastern and western boundary. We therefore are left with term 4 to close the vorticity balance. For southward flow both left-hand side and right-hand side of (6.59) are negative, so term 4 is not required to achieve vorticity balance. In contrast, for northward flow, term 2 and 3 have opposite sign and only a strongly positive term 4 can achieve vorticity balance. $-R\,\partial M_y/\partial x \gg 0$ is, however, only possible at the western boundary. Therefore, friction in the boundary current produces enough positive vorticity that the negative vorticity input by the wind is overcompensated. This enables the movement of the water parcel from south to north.

Henry Stommel examined the deep circulation, as well. He used a similar model which is generally referred to as the Stommel–Arons model presented in two landmark papers (Stommel 1958; Stommel and Arons 1960). These articles led to the remarkable prediction of a western boundary current that is supposed to be located in the Atlantic at a depth of 2–3 km, flowing from north to south! Consequently, physical oceanographers set up an intensive search for this current in order to verify the theoretical prediction. It was finally identified off Cape Hatteras using current meters. Maximum velocities in the core at a depth of 2,500 m are around 20 cm s^{-1}.

At this depth, the effect of the wind can be neglected, however, the mass flux, also a source term in (6.58), must be accounted for. Stommel postulated a large-scale, extremely slow upwelling in the deep ocean in order to compensate for the deep water formation occurring in polar regions. This signifies that water leaves the layer h and hence $Q < 0$ in (6.58). Analogously, Table 6.2 can be compiled.

Table 6.2 Signs of the terms in (6.59) for the Stommel–Arons model on the northern hemisphere. The relation shaded in grey is required in order to close the balance of terms. The large gradients imply a strong, confined flow, i.e., a boundary current.

Direction of flow	1 $\rho h \dfrac{D\zeta}{Dt}$	2 $+\rho h \dfrac{Df}{Dt}$	3 $\approx -(\zeta+f)Q$	4 $-R\dfrac{\partial M_y}{\partial x}$	5 $+R\dfrac{\partial M_x}{\partial y}$
N → S	≈ 0	< 0	> 0	$\ll 0$	≈ 0
S → N	≈ 0	> 0	> 0	≈ 0	≈ 0

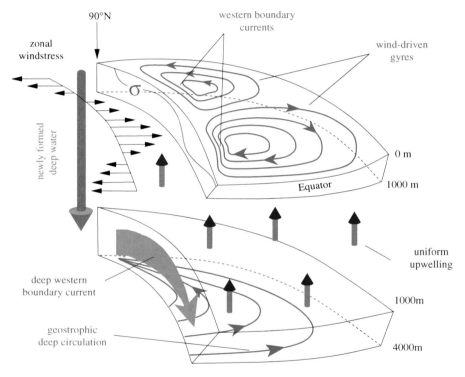

Fig. 6.15 Panoramic and simplified view of the large-scale currents in the northern hemisphere Atlantic based on the Stommel and Stommel–Arons models.

The vorticity balance requires a deep western boundary current flowing southward. It supplies the inner geostrophic flow with water and therefore continuously loses strength. Just this prediction could not be confirmed by observations, which points to a much more complicated picture of deep currents, in particular, the assumption of large-scale uniform upwelling seems inconsistent with recent measurements. A recent critical overview is given in Lozier (2010).

In a highly simplified view, Fig. 6.15 displays the structure of the current systems in the northern hemisphere Atlantic schematically.

Chapter 7
Large-Scale Circulation in the Atmosphere

7.1 Zonal and Meridional Circulation

In this chapter the general circulation in the atmosphere is presented in a simplified form. A comprehensive description of the dynamics of the atmosphere can be found in Holton (2004).

The consideration in Chap. 4 of zonally and temporally averaged quantities and their deviations was useful for the analysis of the meridional heat fluxes. Here, we follow the same approach. Applying suitable time averages the short-term weather events are filtered out and the general circulation can be separated into a quasi-stationary component, a monsoon component that changes its direction during the seasonal cycle, and a component describing low-frequency variations.

The mean flow in the atmosphere is mainly directed from west to east, and so are the highest wind velocities (Figs. 7.1 and 7.2). This is a result of the conservation of the air masses' angular momentum on the rotating Earth. Their movement is driven by the meridional temperature distribution.

The specific angular momentum (angular momentum per mass) of an air parcel that moves along the latitude φ at velocity u relative to the Earth's surface is given by

$$L = (\Omega R \cos\varphi + u) R \cos\varphi, \qquad (7.1)$$

where Ω and R are the angular velocity and the Earth radius, respectively. If no forces act on the air parcel, the angular momentum L is conserved. Consider an air parcel which starts from rest at the equator and reaches latitude φ. Accounting for the conservation of its angular momentum, its zonal velocity reaches

$$u(\varphi) = \frac{\Omega R \sin^2\varphi}{\cos\varphi}. \qquad (7.2)$$

This means, that at 30°N a westerly wind with a velocity of $u = 134 \,\mathrm{m\,s^{-1}}$ would result. This calculation, however, overestimates the speed of the zonal jet stream by about a factor of 3. The observed jet stream maximum is located at 35°N and at an altitude of about 12 km (Fig. 7.2). But this simple computation shows that the

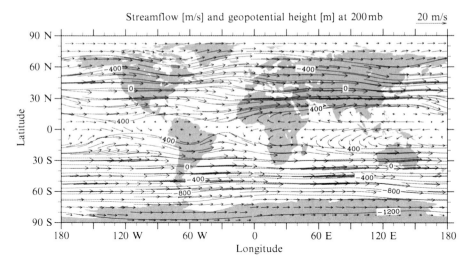

Fig. 7.1 Mean wind field at an altitude of around 12 km. Data from ERA-40 (Uppala et al. 2005). Figure constructed by F. Lehner.

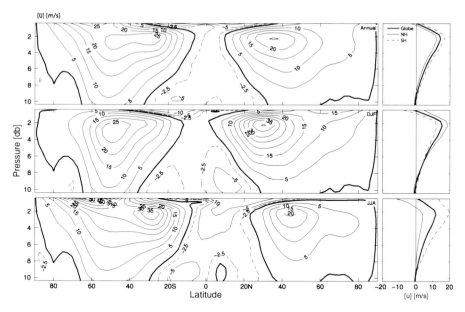

Fig. 7.2 Mean wind in m s^{-1} in a meridional transect of the atmosphere. The strong west-east jets in the northern- and southern hemisphere at an altitude of around 12 km are clearly visible. Data from ERA-40 (Uppala et al. 2005). Figure constructed by F. Lehner.

7.1 Zonal and Meridional Circulation

transport of angular momentum is by far sufficient for an explanation of the high zonal wind velocities at mid-latitudes. However, it also leads to the conclusion that angular momentum must be constantly removed from the flow. This is caused by eddies and the associated transport of angular momentum. The mean meridional advective transport of angular momentum is given by

$$[\overline{vL}] = \underbrace{[\bar{v}]\,(\Omega R \cos\varphi + [\bar{u}])\,R\cos\varphi}_{M} + (\underbrace{[\overline{\bar{u}^*\bar{v}^*}]}_{SE} + \underbrace{[\overline{u'v'}]}_{TE})\,R\cos\varphi\,, \quad (7.3)$$

in analogy to (4.7). The meridional transport of angular momentum is achieved by the combination of the mean flow (M), stationary eddies (SE) and transient eddies (TE). Observations show that at latitudes between 20° and 50° TE is the largest contribution to angular momentum transport.

In the eighteenth century, *George Hadley* proposed that the strong solar radiation in the tropics heats up the air and causes it to rise. On the northern hemisphere the resulting near-surface flow is directed towards the equator and converges finally at the so-called intertropical convergence zone (ITCZ). Its deviation towards the west (so that the zonal velocity is westward, $u < 0$) is a result of angular momentum conservation. This causes the well-known *trade winds*. The return flow at higher levels is analogously deviated towards the east ($u > 0$) inducing a zonal jet stream at higher latitudes where it passes over to descending air motions. The resulting meridionally closed circulation is referred to as *Hadley circulation* or *Hadley cell*, schematically depicted in Fig. 7.3.

The effect of the Coriolis force, or of the conservation of angular momentum respectively, is hence a south-west-directed flow at the surface and a north-east-directed flow at high altitudes. Hadley expected the circulation cell to extend all the way to the pole. However, observations indicated that the Hadley cell does not even reach the mid-latitudes, because there, the mean winds are directed to the east at the surface, as well as at high altitudes (westerlies). The simple picture of a merely thermally-driven flow is therefore not sufficient to explain observations outside the tropics.

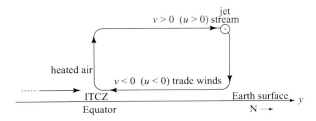

Fig. 7.3 A simple depiction of the Hadley circulation in the northern hemisphere. Heated air at the equator rises first, then moves polewards, descends at higher latitudes and finally returns to the equator as a near-surface flow.

For a deeper understanding the balance equations for momentum, mass and energy in the atmosphere need to be solved. The equation of motion is basically analogous to (6.13) and the continuity equation is given with (3.7); in addition, the thermodynamic energy equation must be taken into account. For a complete derivation of the equations, the reader is referred to Holton (2004).

We consider the zonally and temporally averaged equations, where terms of the form (4.6) will occur. The flow in a meridional plane can be described by a meridional stream function $\chi(y, z)$, defined as follows:

$$\rho_0 \bar{v} = -\frac{\partial \bar{\chi}}{\partial z} \tag{7.4a}$$

$$\rho_0 \bar{w} = \frac{\partial \bar{\chi}}{\partial y}, \tag{7.4b}$$

where v and w are meridional and vertical velocities and $\rho_0 = \rho_0(z)$ is the density of air. The overbars denote appropriate time averaging. As derived in Holton (2004), the stream function satisfies the following partial differential equation:

$$\frac{N^2}{\rho_0} \frac{\partial^2 \bar{\chi}}{\partial y^2} + f_0^2 \frac{\partial}{\partial z}\left(\frac{1}{\rho_0} \frac{\partial \bar{\chi}}{\partial z}\right) = \underbrace{\frac{\kappa}{H} \frac{\partial \bar{J}}{\partial y}}_{D} - \underbrace{\frac{R^*}{H} \frac{\partial^2 \overline{v'T'}}{\partial y^2}}_{TEH} - \underbrace{f_0 \frac{\partial^2 \overline{v'u'}}{\partial z \partial y}}_{TEM} + \underbrace{f_0 \frac{\partial \bar{X}}{\partial z}}_{R}.$$
(7.5)

Here, N is the Brunt–Väisälä frequency, the angular frequency of free vertical oscillations in a stable atmosphere given by

$$N^2 = \frac{R^*}{H}\left(\frac{\kappa T_0}{H} + \frac{dT_0}{dz}\right) \tag{7.6}$$

(which is approximately constant in the troposphere), where R^* is the specific gas constant of the air and $\kappa = R^*/c_p$; furthermore, $H = R^* \hat{T}_0/g$ is the isothermal scale-height of the atmospheric layer considered here with temperature $T_0 = T_0(z)$ and a layer mean temperature $\hat{T}_0$. The physical quantity $\bar{J}(y, z)$ in (7.5) is a mean diabatic heating rate (induced by heat fluxes at the ground or latent heat from condensation processes) and $\bar{X}$ is a mean drag in a zonal direction by friction at the ground. Finally, the coordinate z in (7.5) signifies the so-called log-pressure coordinate $z = -H \ln(p/p_s)$ with p_s the air pressure on the underside of the layer. In the troposphere, the log-pressure coordinate is nearly equal to the usual z-coordinate which represents a geometric height coordinate. According to (7.5), the stream function is driven by four processes: (a), diabatic heat sources (D), (b), heat fluxes associated with transient eddies (TEH), (c), fluxes of momentum associated with transient eddies (TEM) and, (d), friction (R).

Equation (7.5) is a generalized form of the Poisson equation and needs to be complemented by boundary conditions. Therefore, we consider a domain, reaching from the equator nearly to the pole and in the vertical dimension from the Earth surface

7.1 Zonal and Meridional Circulation

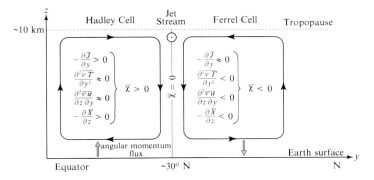

Fig. 7.4 Schematic illustration of the stream function $\overline{\chi}$ in the northern hemisphere. $\overline{\chi} > 0$ is the thermally direct Hadley cell, $\overline{\chi} < 0$ describes the thermally indirect Ferrel cell. Angular momentum is supplied to the atmosphere south of about 30°N and removed from the atmosphere north of it.

up to the tropopause. Transport is assumed to be confined within these boundaries and hence $\overline{\chi} = 0$ on the boundary. The domain is illustrated in Fig. 7.4.

For the qualitative discussion of (7.5), we assume that $\overline{\chi}$ can be represented by appropriate sin-functions in y and z which satisfy the boundary conditions. Hence, the left-hand side of (7.5) is proportional to $-\overline{\chi}$ and we can derive the following relations:

$$\overline{\chi} \propto -\frac{\partial}{\partial y} \text{ (diabatic heat sources)} + \frac{\partial^2}{\partial y^2} \text{ (meridional eddy heat flux)}$$

$$+ \frac{\partial^2}{\partial z \partial y} \text{ (meridional eddy momentum flux)} - \frac{\partial}{\partial z} \text{ (zonal shear)}. \quad (7.7)$$

Close to the equator, a large amount of latent heat is released and hence, $\overline{J} > 0$, while at around 30°N and further to the north cooling caused by radiative losses dominates, hence $\overline{J} < 0$. Between the equator and 30°N $\overline{J}$ decreases and hence $\partial \overline{J}/\partial y < 0$. In these latitudes the eddy fluxes TEH and TEM are small; their contribution to the zonal wind stress, that is directed towards the east due to the trade winds, is only to be considered at its lower boundary. Term D prevails in (7.5) and contributes, together with the smaller term R, to the observed Hadley cell, a meridional cell with $\overline{\chi} > 0$. This is denoted as a *thermally direct* cell, i.e., warm air rises, while colder air sinks (Fig. 7.4).

The eddy activity has a maximum at around 30°–60°N where the storm tracks are located. The latitudinal and altitudinal dependence of the meridional eddy fluxes are illustrated in Fig. 7.5. It can be shown that at these latitudes the two respective terms are negative in (7.7). Due to the westerlies, the drag is directed towards the west and decreases in magnitude with increasing altitude, hence $-f_0 \partial \overline{X}/\partial z < 0$. Therefore, according to (7.7), $\overline{\chi} < 0$ and an *indirect* cell is formed. This indirect cell in the region of 40°–60°N is called *Ferrel cell* (Fig. 7.4). The Ferrel cell is thermally indirect, i.e., cold air rises and warm air sinks.

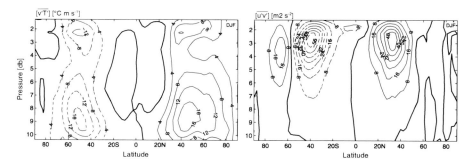

Fig. 7.5 Observed distribution of eddy fluxes of heat (**left**, in $°C\,m\,s^{-1}$) and momentum (**right**, in $m^2\,s^{-2}$) for northern winter. Positive fluxes are directed northward. Data from ERA-40 (Uppala et al. 2005). Figure constructed by F. Lehner.

A part of the specific angular momentum (7.1) of the northern hemisphere is produced in the Hadley cell in the region of the trade winds, where $u < 0$. Here, the air is accelerated by friction at the Earth surface so that a flux of angular momentum from the Earth to the atmosphere is induced (Fig. 7.4). This angular momentum is transported polewards to the Ferrel cell and subsequently again lost to the Earth surface in mid-latitudes, where $u > 0$.

The observed meridional circulation (Fig. 7.6) shows strong Hadley cells in the respective winter hemisphere. The Ferrel cells in the southern and northern hemispheres can also be identified. The simplified theoretical model in (7.5) captures this structure quite well.

7.2 The Lorenz–Saltzman Model

In order to examine the thermally-driven flow, Barry Saltzman (1931–2001) derived an approximation consisting of a non-linear system of ordinary differential equations from the governing equations of a viscous, stably stratified flow (Saltzman 1962). The fundamental significance of this equation system was recognized by Edward Lorenz who numerically solved this system and interpreted it (Lorenz 1963). Beyond the particular application for viscous incompressible fluids, the system may be interpreted as the simplest form of a description of non-linear processes in relation with the general circulation in the atmosphere. The model is of particular significance because it was the first system to describe deterministic chaos and, based on it, *Chaos theory* was developed.

Deterministic chaos can occur in a non-linear system (non-linearity is a necessary but not satisfactory condition) and is based on the fact that the instantaneous time derivative is given functionally, however, the temporal evolution of the system cannot be predicted over long periods. Mathematically speaking, the system is determined by several coupled ordinary differential equations of first order in time. Its

7.2 The Lorenz–Saltzman Model

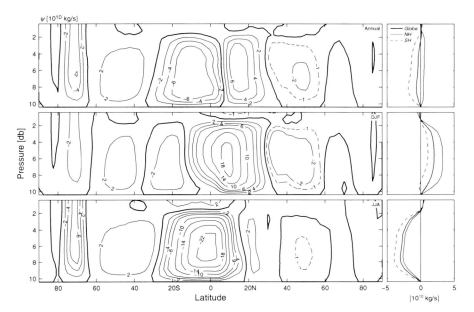

Fig. 7.6 Observed meridional circulation (stream function in 10^{10} kg s^{-1}), annually averaged (**upper**), for the northern winter (**middle**) and the northern summer (**lower**). Data from ERA-40 (Uppala et al. 2005). Figure constructed by F. Lehner.

changes can be calculated exactly at all times: the system is therefore deterministic. This system is generally referred to as the Lorenz model. But since the original equations were derived by B. Saltzman, we shall call it *Lorenz–Saltzman model*.

The following derivation of the Lorenz–Saltzman model is somewhat technical. Nevertheless, it will be described here, since in the literature only the dimensionless system is usually given. The Lorenz–Saltzman model is formulated on a meridional plane in the non-rotating reference system (y, z). A generalization for the f-plane was realized later (Lorenz 1984). Solutions are assumed uniform in the x-direction. We further assume, that diabatic effects, e.g., heat sources, drive the flow clock-wise. Additionally, a constant vertical temperature gradient is chosen as a background state. The solution domain is shown schematically in Fig. 7.7.

The fluid is considered incompressible (in fact, not a valid approximation for the atmosphere, but applicable to a water body), therefore, mass conservation is given by the continuity equation (6.14),

$$\frac{\partial v}{\partial y} + \frac{\partial w}{\partial z} = 0. \tag{7.8}$$

With this, a stream function can be defined as follows:

$$v = -\frac{\partial \Psi}{\partial z}, \qquad w = \frac{\partial \Psi}{\partial y}. \tag{7.9}$$

Fig. 7.7 Coordinates and solution domain for the Lorenz–Saltzman model. A constant vertical temperature gradient is chosen.

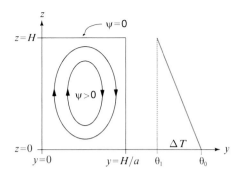

Thus, the relative vorticity in the meridional y-z-plane, i.e. $\zeta = \partial w/\partial y - \partial v/\partial z$, is given by

$$\zeta = \vec{\nabla}^2 \Psi. \tag{7.10}$$

The formulation of the conservation equation of vorticity reveals that vorticity is dissipated by molecular diffusion and produced by meridional temperature gradients $\partial \theta/\partial y$ (buoyancy). In order to derive the vorticity equation, we start from the momentum equations

$$\frac{Dv}{Dt} = -\frac{1}{\rho_0} \frac{\partial p}{\partial y} + \nu \vec{\nabla}^2 v \tag{7.11a}$$

$$\frac{Dw}{Dt} = -\frac{1}{\rho_0} \frac{\partial p}{\partial z} + \nu \vec{\nabla}^2 w - \frac{g}{\rho_0} \tilde{\rho}, \tag{7.11b}$$

where ν is the kinematic viscosity and the last term in (7.11b) describes the acceleration due to buoyancy, caused by a small deviation $\tilde{\rho}$ from the constant density ρ_0 (Archimedes' principle). Cross-differentiating $\partial(7.11b)/\partial y - \partial(7.11a)/\partial z$ and considering (7.8) yields

$$\frac{D\zeta}{Dt} = \nu \vec{\nabla}^2 \zeta - \frac{g}{\rho_0} \frac{\partial \tilde{\rho}}{\partial y}. \tag{7.12}$$

Using the volume coefficient of expansion

$$\alpha = -\frac{1}{\rho_0} \frac{\partial \tilde{\rho}}{\partial \theta}, \tag{7.13}$$

(7.12) can be rewritten as

$$\frac{D\zeta}{Dt} = \nu \vec{\nabla}^2 \zeta + g \alpha \frac{\partial \theta}{\partial y},$$

i.e.,

$$\frac{\partial \zeta}{\partial t} + v \frac{\partial \zeta}{\partial y} + w \frac{\partial \zeta}{\partial z} = \nu \vec{\nabla}^2 \zeta + g \alpha \frac{\partial \theta}{\partial y}. \tag{7.14}$$

7.2 The Lorenz–Saltzman Model

We now assume the following temperature distribution

$$\theta(y,z,t) = \theta_0 - \frac{\Delta T}{H}z + \tilde{\theta}(y,z,t), \qquad (7.15)$$

where $\tilde{\theta}$ is the deviation from a stable linear temperature profile $\theta_0 - \Delta T/H z$ with $\Delta T = \theta_0 - \theta_1$ (Fig. 7.7). The conservation of thermal energy can be captured by the heat equation

$$\frac{D\theta}{Dt} = \kappa \vec{\nabla}^2 \theta; \qquad (7.16)$$

considering (7.15), we obtain

$$\frac{\partial \tilde{\theta}}{\partial t} + v\frac{\partial \tilde{\theta}}{\partial y} - w\frac{\Delta T}{H} + w\frac{\partial \tilde{\theta}}{\partial z} = \kappa \frac{\partial^2 \tilde{\theta}}{\partial y^2} + \kappa \frac{\partial^2 \tilde{\theta}}{\partial z^2}. \qquad (7.17)$$

Here, κ is the thermal diffusivity. Inserting (7.9) and (7.10) into (7.14) and (7.17) results with (7.15) in the following system:

$$\frac{\partial}{\partial t}\vec{\nabla}^2\Psi - \frac{\partial \Psi}{\partial z}\frac{\partial}{\partial y}\vec{\nabla}^2\Psi + \frac{\partial \Psi}{\partial y}\frac{\partial}{\partial z}\vec{\nabla}^2\Psi = \nu \vec{\nabla}^4 \Psi + g\alpha \frac{\partial \tilde{\theta}}{\partial y} \qquad (7.18)$$

$$\frac{\partial \tilde{\theta}}{\partial t} - \frac{\partial \Psi}{\partial z}\frac{\partial \tilde{\theta}}{\partial y} + \frac{\partial \Psi}{\partial y}\frac{\partial \tilde{\theta}}{\partial z} = \kappa \vec{\nabla}^2 \tilde{\theta} + \frac{\Delta T}{H}\frac{\partial \Psi}{\partial y}. \qquad (7.19)$$

Equations (7.18) and (7.19) represent a coupled, non-linear system of partial differential equations which has to be completed by boundary conditions. We postulate no transport across the boundaries and no heat flux across the meridional boundaries. Furthermore, fixed temperatures at the ground and at the upper boundary shall be given, hence

$$\Psi = 0 \quad \text{at the boundary}, \qquad (7.20a)$$

$$\frac{\partial \tilde{\theta}}{\partial y} = 0 \quad \text{for } y = 0 \text{ and } y = H/a, \qquad (7.20b)$$

$$\tilde{\theta} = 0 \quad \text{for } z = 0 \text{ and } z = H. \qquad (7.20c)$$

The solution of this system is supposed to be found approximately by only considering the rough spatial structure inside the solution domain. To do so, we assume a truncated *Fourier expansion* satisfying the boundary conditions:

$$\Psi(y,z,t) = X(t)\sin\left(\frac{\pi a y}{H}\right)\sin\left(\frac{\pi z}{H}\right) \qquad (7.21a)$$

$$\tilde{\theta}(y,z,t) = Y(t)\cos\left(\frac{\pi a y}{H}\right)\sin\left(\frac{\pi z}{H}\right) - Z(t)\sin\left(\frac{2\pi z}{H}\right). \qquad (7.21b)$$

The space dependence is prescribed, the time dependence is given by the coefficient functions $X(t)$, $Y(t)$ and $Z(t)$. This *a priori* choice allows solutions with the simplest possible structure and, due to the truncation of the expansion only represents approximate solutions. Inserting (7.21) into (7.18), and eliminating the common factor $\sin(\pi a y/H) \sin(\pi z/H)$ we find

$$\left(\frac{\pi}{H}\right)^2 (1+a^2) \frac{dX}{dt} = -\nu \left(\frac{\pi}{H}\right)^4 (1+a^2)^2 X + g\alpha \frac{\pi a}{H} Y. \tag{7.22}$$

Similarly, inserting (7.21) into (7.19) yields

$$\cos\left(\frac{\pi a y}{H}\right) \sin\left(\frac{\pi z}{H}\right) \left\{ \frac{dY}{dt} - \frac{\pi a}{H} \frac{2\pi}{H} XZ \cos\left(\frac{2\pi z}{H}\right) \right.$$

$$\left. + \kappa \left(\frac{\pi}{H}\right)^2 (1+a^2) Y - \frac{\Delta T}{H} \frac{\pi a}{H} X \right\}$$

$$= \sin\left(\frac{2\pi z}{H}\right) \left\{ \frac{dZ}{dt} - \frac{1}{2} \frac{\pi a}{H} \frac{\pi}{H} XY + \kappa \left(\frac{2\pi}{H}\right)^2 Z \right\},$$

i.e., with $\sin\left(\frac{2\pi z}{H}\right) = 2 \sin\left(\frac{\pi z}{H}\right) \cos\left(\frac{\pi z}{H}\right)$,

$$\cos\left(\frac{\pi a y}{H}\right) \left\{ \frac{dY}{dt} - \frac{\pi a}{H} \frac{2\pi}{H} XZ \cos\left(\frac{2\pi z}{H}\right) \right.$$

$$\left. + \kappa \left(\frac{\pi}{H}\right)^2 (1+a^2) Y - \frac{\Delta T}{H} \frac{\pi a}{H} X \right\}$$

$$= 2 \cos\left(\frac{\pi z}{H}\right) \left\{ \frac{dZ}{dt} - \frac{1}{2} \frac{\pi a}{H} \frac{\pi}{H} XY + \kappa \left(\frac{2\pi}{H}\right)^2 Z \right\}. \tag{7.23}$$

Since this equation has to be valid for all values $0 \leq y \leq H/a$ and $0 \leq z \leq H$, the sums in the two {}-brackets have to vanish. Finally, we assume that the dynamics are determined by processes inside the vertical range $1/4\, H < z < 3/4\, H$ and hence, the rough approximation $\cos(2\pi z/H) \approx -1$ is applicable.

The system of ordinary differential equations for the coefficient functions $X(t)$, $Y(t)$ and $Z(t)$ reads:

$$\frac{dX}{dt} = -cX + dY, \tag{7.24a}$$

$$\frac{dY}{dt} = -eXZ + fX - gY, \tag{7.24b}$$

$$\frac{dZ}{dt} = hXY - kZ, \tag{7.24c}$$

7.2 The Lorenz–Saltzman Model

with the seven constants

$$c = \nu \left(\frac{\pi}{H}\right)^2 (1+a^2), \quad d = \frac{g\alpha a H}{\pi(1+a^2)},$$

$$e = \frac{2\pi^2 a}{H^2}, \quad f = \frac{\Delta T \pi a}{H^2}, \quad g = \kappa \left(\frac{\pi}{H}\right)^2 (1+a^2), \quad (7.25)$$

$$h = \frac{\pi^2 a}{2H^2}, \quad k = 4\kappa \left(\frac{\pi}{H}\right)^2.$$

By introducing new dimensionless physical quantities t, X, Y and Z in the following way,

$$(\frac{\pi}{H})^2 (1+a^2) \kappa t \to t$$

$$\frac{a}{\kappa(1+a^2)} X \to X$$

$$\frac{a}{\kappa(1+a^2)} \frac{g\alpha a H^3}{\pi^3 (1+a^2)^2 \nu} Y \to Y$$

$$2 \frac{a}{\kappa(1+a^2)} \frac{g\alpha a H^3}{\pi^3 (1+a^2)^2 \nu} Z \to Z,$$

the classical Lorenz–Saltzman model can be derived:

$$\frac{dX}{dt} = -\sigma X + \sigma Y \quad (7.26a)$$

$$\frac{dY}{dt} = -XZ + rX - Y \quad (7.26b)$$

$$\frac{dZ}{dt} = XY - bZ \quad (7.26c)$$

with

$$\sigma = \frac{\nu}{\kappa}, \quad r = \frac{g\alpha H^3 \Delta T}{\nu \kappa} \frac{a^2}{\pi^4 (1+a^2)^3}, \quad b = \frac{4}{1+a^2}. \quad (7.27)$$

Note, the quantities t, X, Y and Z in (7.26) are the scaled forms of the quantities t, X, Y and Z in (7.24); for simplicity we do not introduce a new notation.

Figures 7.8 and 7.9 illustrate the solution of (7.26) for a given set of parameters. The time series exhibit a chaotic behaviour, where the variables, here $Y(t)$, change from one regime ($Y > 0$) to the other ($Y < 0$) in an irregular way (Fig. 7.8). The residence time in a certain regime is erratic and considerably longer than the transition between the regimes itself. The system obviously evolves on two different

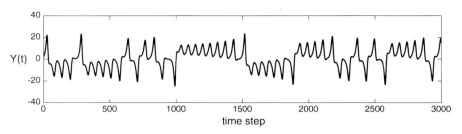

Fig. 7.8 First 3,000 time steps of the time series $Y(t)$ of the Lorenz–Saltzman model with the classical parameter values $r = 28$, $\sigma = 10$, $b = 8/3$ and $\Delta t = 0.012$, integrated using the Runge–Kutta scheme.

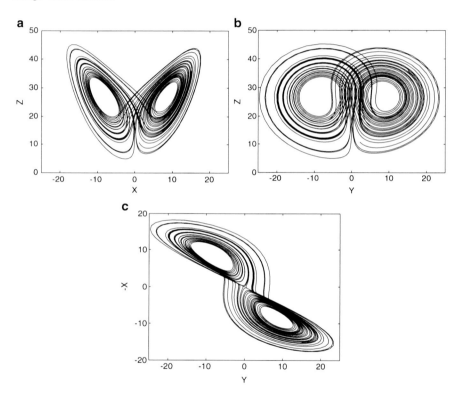

Fig. 7.9 Cross-sections through the different planes in the (X, Y, Z)-space of the Lorenz–Saltzman model with the classical parameter values (see Fig. 7.8). (**a**) (X, Z)-plane. (**b**) (Y, Z)-plane. (**c**) (X, Y)-plane.

time scales: one for the transition and one for the residence time in one regime. The Lorenz–Saltzman model is a prime example for abrupt changes in a dynamical system. These transitions are not a response to external disturbances, but are *spontaneously* triggered by the dynamics of the system itself.

7.2 The Lorenz–Saltzman Model

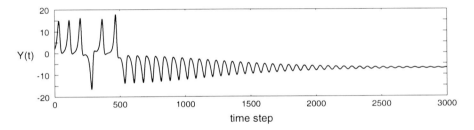

Fig. 7.10 As Fig. 7.8 but with $r = 20$, $\sigma = 10$, $b = 8/3$.

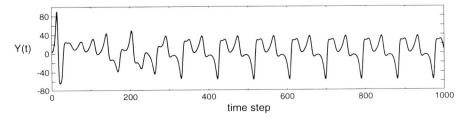

Fig. 7.11 As Fig. 7.8 but with $r = 100$, $\sigma = 10$, $b = 8/3$.

Figure 7.9 shows the trajectories of the Lorenz model at subsequent time steps in the three-dimensional variable space (X, Y, Z). The points $(X = \pm 8.49$, $Y = \pm 8.49$, $Z = 27)$ represent unstable equilibria. Trajectories originating in their surroundings move away from these points in spirals. For chaotic behaviour, as is the case in Figs. 7.8 and 7.9, the trajectories will never cross in the (X, Y, Z)-space. Another particular point is the origin $(0, 0, 0)$, representing another stationary solution of the equations (7.26). It is located in the center of the "transition point" from one regime to the other and hence, is the location of highest "non-predictability" in the Lorenz–Saltzman system.

It is remarkable that this system can exhibit chaotic, periodic or stationary behaviour depending on the choice of parameters (7.27). The chaotic behaviour of the Lorenz–Saltzman model only occurs in certain "windows" of parameter values. Outside these windows, either a stable equilibrium is reached after a relatively long transient phase (Fig. 7.10) or a periodic behaviour can be observed. These are *self-sustained oscillations*, as shown in Fig. 7.11. They develop after a transient phase.

Chapter 8
Atmosphere–Ocean Interactions

8.1 Coupling of Physical Model Components

Energy, momentum and matter (water, carbon, nitrogen, ...) are exchanged between the ocean and the atmosphere. Most of the movements in the ocean, particularly the large-scale flow, are caused by these exchange fluxes. Consequently, they need to be reproduced in a climate model as realistically as possible. In the context of this book we will not treat micro-scale fluxes, occurring on a cm- or smaller scale. An in-depth description is provided by Kraus and Businger (1994). We will only present the parameterisations that are implemented mainly in climate models of coarse resolution. Formulations of so-called *boundary layers* in the atmosphere and ocean are also not discussed.

In the present chapter, we consider primarily heat fluxes (fluxes of thermal or latent energy), water fluxes and momentum fluxes (Fig. 8.1). They are influenced by the dynamics of the atmosphere and the ocean whilst they influence these dynamics. For the individual model components, the fluxes can be considered and formulated as boundary conditions.

Similar considerations have to be made for the coupling of sea ice, ice sheets, and land surface modules.

8.2 Thermal Boundary Conditions

The complete thermal boundary condition for the heat flux from the ocean to the atmosphere $F^{O \rightarrow A}$ is given by

$$F^{O \rightarrow A} = \underbrace{-(1-\alpha_O) \, Q^{\text{short}}}_{\text{SW}} + \underbrace{\varepsilon_O \, \sigma \, T_O^4}_{\text{LW}} \underbrace{- \varepsilon_A \, \sigma \, T_A^4}_{\text{LB}} + \underbrace{D \, (T_O - T_A)}_{\text{S}} + \underbrace{E(T_O, T_A)}_{\text{E}}, \tag{8.1}$$

where T_O and T_A are the surface temperatures of the ocean and the atmosphere, respectively, Q^{short} denotes the (mainly short-wave) solar radiation impinging on the ocean surface, α_O the albedo of the ocean surface, ε_O and ε_A the emissivities

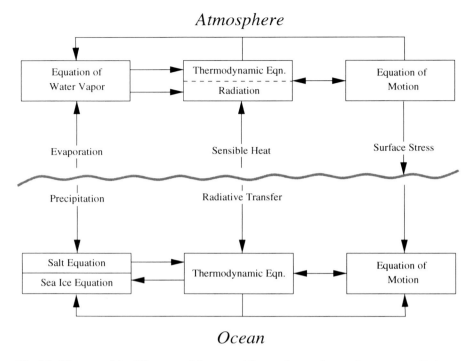

Fig. 8.1 Diagram of the different model parts and fluxes of water, heat and momentum. Redrawn from Fig. 1 in Manabe and Stouffer (1988).

of the ocean surface and the atmosphere, respectively, D a transfer coefficient for the sensible heat flux and finally $E(T_O, T_A)$ a relation describing the evaporation on the ocean surface. The heat flux consists of five components: short-wave solar radiation (SW), long-wave radiation of the ocean (LW), long-wave back radiation of the atmosphere (LB), sensible heat flux (S) and evaporation (E). A positive sign denotes a flux from the ocean to the atmosphere. The global distribution of ocean-atmosphere fluxes is given in Fig. 8.2 and for the Atlantic in Fig. 8.3.

Typical values for the parameters in (8.1) are

$$\alpha_O = 0.2, \quad \varepsilon_O = 0.96, \quad \varepsilon_A = 0.7 \ldots 0.9, \quad D = 10 \, \text{W} \, \text{K}^{-1} \, \text{m}^{-2}. \quad (8.2)$$

For certain parameterisations, the transfer coefficient D for the sensible heat flux may depend on wind speed.

The temperature dependence of the evaporation E can be expressed as a Taylor series expanded about the temperature of the atmosphere T_A (Haney 1971),

$$E(T_O, T_A) = E(T_O = T_A, T_A) + \left.\frac{\mathrm{d}E(T_O, T_A)}{\mathrm{d}T_O}\right|_{T_O = T_A} (T_O - T_A) + \ldots .$$

8.2 Thermal Boundary Conditions

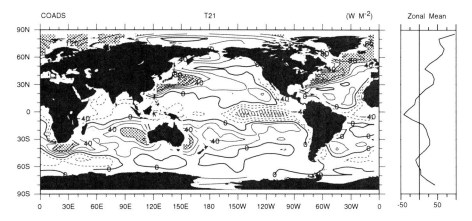

Fig. 8.2 Left: Annually averaged heat fluxes in W m^{-2} based on the *Comprehensive Ocean Atmosphere Data Set* (COADS), Woodruff et al. (1987). Areas with heat fluxes exceeding 60 W m^{-2} are hatched. **Right**: Zonal average. Figure from Trenberth et al. (2001).

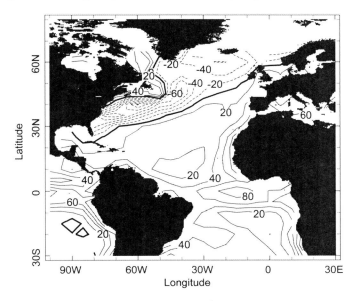

Fig. 8.3 Heat flux $-F^{O \to A}$ in the Atlantic in W m^{-2}. The map was compiled on the internet (http://ingrid.ldeo.columbia.edu), where many data sets are available. Here we have used the Cayan data set, Cayan (1992).

An appropriate linear truncated Taylor series, which is in accordance with the Clausius–Clapeyron equation, reads (Gill 1982; Stocker et al. 1992)

$$E(T_O, T_A) = c_E \, e^{14.7 - \frac{5418 \, K}{T_A}} \left(0.2 + 5418 \, K \, \frac{T_O - T_A}{T_A^2} \right), \quad (8.3)$$

where c_E is a transfer coefficient depending on the wind speed.

For simplicity, in (8.1), the long-wave heat fluxes are given as grey body radiation with their associated emissivities. However, particularly LB may originate from the individual contributions of the free atmosphere and the reflection of clouds (high, as well as low clouds) and hence, may depend on modeled variables of the atmosphere component in a complex way. They also affect the solar radiation Q^{short} which is prescribed in (8.1) but in reality this also depends on the state of the atmosphere. In principle, the heat flux between ocean and atmosphere depends on the temperatures in both components as well as on the wind speeds. These are all quantities that are simulated in a coupled climate model.

In climate modelling, simplified forms of (8.1) are often applied, especially when a single model component (e.g., the ocean) is integrated individually or in models of reduced complexity. This is often the case at the beginning of a simulation, when a stable equilibrium climate has to be reached. An adequately simplified form of (8.1) for an ocean model is found by linearizing this relation using a truncated Taylor series about the temperature of the atmosphere T_A, which is assumed to be constant (for an atmospheric model analogously):

$$F^{O \to A}(T_O) = F^{O \to A}(T_A) + \frac{dF^{O \to A}}{dT_O}\bigg|_{T_O=T_A} (T_O - T_A)$$

$$= F_0 + D^* (T_O - T_A), \qquad (8.4)$$

where

$$F_0 = F^{O \to A}(T_A),$$

$$D^* = \frac{dF^{O \to A}}{dT_O}\bigg|_{T_O=T_A}.$$

F_0 is the net heat flux through the ocean surface of the temperature $T_O = T_A$ and $D^* \approx 45\,\text{W}\,\text{K}^{-1}\,\text{m}^{-2}$ is a typical transfer coefficient. Note, that $D^* > D$, since (8.4) contains the effects of temperature-dependence of evaporation and of the net long-wave radiation. Haney (1971) proposed a further simplification of (8.4),

$$F^{O \to A}(T_O) = D^* (T_O - T_O^*), \qquad (8.5)$$

with the so-called *restoring temperature*

$$T_O^* = T_A - \frac{F_0}{D^*}, \qquad (8.6)$$

which is assumed to be constant. The formulation (8.5) is called *restoring heat flux* or *Newtonian heat flux*. This is due to the fact that heat fluxes are directed in a way that the variable surface temperature T_O asymptotically approximates the fixed temperature T_O^* when no other heat fluxes (e.g., advective heat fluxes) are present.

8.2 Thermal Boundary Conditions

Fig. 8.4 1-box model for the illustration of restoring fluxes.

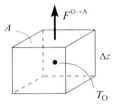

The effect of the restoring heat fluxes and the role of D^* shall be briefly illustrated by means of a 1-box model. The energy balance in the 1-box model (Fig. 8.4) with surface area A and volume $V = A\,\Delta z$ is given by

$$\rho V c \frac{dT_O}{dt} = -A\, F^{O\to A}(T_O) = -A\, D^*\left(T_O - T_O^*\right). \tag{8.7}$$

ρ is the mass density and c the specific heat capacity of the ocean water. Equation (8.7) can be written as follows:

$$\frac{d\left(T_O - T_O^*\right)}{dt} = -\frac{D^*}{\rho c\,\Delta z}\left(T_O - T_O^*\right)$$

and the solution is

$$T_O(t) = T_O^* + \left(T_O(t=0) - T_O^*\right) e^{-\frac{t}{\tau}},$$

where τ is a typical time scale of the relaxation. Accordingly, disturbances decay on a time scale τ. An estimate of this *restoring time scale* yields:

$$\tau = \frac{\rho c\,\Delta z}{D^*} \approx \frac{1028 \cdot 3900}{45}\,\frac{s}{m}\,\Delta z \approx 1\,\frac{day}{m}\,\Delta z. \tag{8.8}$$

In an ocean model used in an uncoupled mode, *restoring heat fluxes* are commonly used and for the *restoring temperature* T_O^* the observed surface temperature is applied. This guarantees that the surface values of temperature never deviate too far from the observations and that a defined equilibrium of the currents is reached. In atmosphere models, one can proceed analogously.

The formulation (8.5) as part of an ocean model may be regarded as the simplest form of a specific parameterisation of the effect of the not dynamically modeled atmosphere. A closer investigation of two extreme cases of ocean models demonstrates this:

- Constant temperature of the atmosphere T_A: The atmosphere acts as a "heat reservoir" with infinite heat capacity. These are *infinite heat capacity models*.
- Constant flux from the ocean to the atmosphere: We select $D^* = 0$ in (8.4), implying that the heat flux $F^{O\to A}(T_O)$ is constant, $F^{O\to A}(T_O) = F_0$, and therefore independent of a possible deviation from the mean temperatures of ocean

and atmosphere. The heat capacity of the atmosphere vanishes, the atmosphere radiates the heat energy immediately to space. This is referred to as *zero heat capacity models*, and is also the case for very long relaxation times.

Analogous considerations apply for atmospheric models.

The fact that (8.5) ignores a possible scale-dependence of the relaxation time is an important problem. Small-scale temperature anomalies at the sea surface are eliminated at a rate of ≈ 1 m per day by direct heat exchange. But large-scale anomalies may persist much longer, since they penetrate deeper into the ocean, and hence require a significantly higher amount of heat for equilibration. Under certain circumstances this heat energy cannot be provided by the atmosphere (e.g., in the form of rapidly passing storms).

In order to account for the scale-dependence, we would have to write

$$F(\vec{x}) = \int \Lambda(\vec{x}, \vec{x}\,') \{T(\vec{x}\,') - T^*(\vec{x}\,')\}\, d\vec{x}\,', \qquad (8.9)$$

which contains a non-local dependence of the fluxes at location $\vec{x}$. Here, the determination of the form function $\Lambda(\vec{x}_1, \vec{x}_2)$ is a challenge. A step towards scale dependence of τ or D was proposed by Willebrand (1993),

$$F^{O \to A}(T_O) = D_1 \left(T_O - T_O^*\right) - D_2 \vec{\nabla}^2 \left(T_O - T_O^*\right), \qquad (8.10)$$

with $D_1 \approx 2\,\mathrm{W\,K^{-1}\,m^{-2}}$ and $D_2 \approx 10^{13}\,\mathrm{W\,K^{-1}}$. Small-scale anomalies at a typical spatial scale of 500 km, given a surface ocean layer of 50 m thickness, are equilibrated on a time scale of $\tau_2 = \rho c\,\Delta z\,L^2/D_2 \approx 60$ days, while large-scale anomalies decay on a time scale of $\tau_1 = \rho c\,\Delta z/D_1 \approx 3.5$ years. Formulation (8.10) may also be interpreted as a compact form of an atmospheric energy balance model.

8.3 Hydrological Boundary Conditions

The coupling of the water cycle is of fundamental significance for the transport of energy in the form of latent heat in the atmosphere and for the change in density of the surface water, induced by precipitation and evaporation. Evaporation separates water and salt and only the latter remains in the ocean.

Accordingly, evaporation leads to an increase of the salinity in the surface ocean. The density of sea water at the surface $\rho(T, S)$ can be expressed as a Taylor series expanded about a temperature T_0 and a salinity S_0. An appropriately truncated Taylor series reads

$$\rho(T, S) = \rho_0 \left(1 + \alpha\,(T - T_0) + \beta\,(S - S_0) + \gamma\,(T - T_0)^2\right), \qquad (8.11)$$

8.3 Hydrological Boundary Conditions

where

$$\rho_0 = 1028 \text{ kg m}^{-3} \quad \alpha = -5.4128 \cdot 10^{-5} \text{ K}^{-1}$$
$$T_0 = 0°C \quad \beta = 7.623 \cdot 10^{-4} \quad (8.12)$$
$$S_0 = 35 \quad \gamma = -5.0804 \cdot 10^{-6} \text{ K}^{-2}.$$

ρ decreases with increasing temperature T and increases with increasing salinity S (S in g salt per kg water).

In analogy to (8.5), ocean models are run to equilibrium with the boundary condition

$$F_S^{O \to A}(S) = D_S^* \left(S - S^* \right) \quad (8.13)$$

(the transfer coefficient D_S^* has the units kg m^{-2} s^{-1}). This guarantees surface salinity values to remain close to the observational data S^*. Formulation (8.13) is called *restoring salt flux*. Here, the *restoring time scale*

$$\tau = \frac{\rho \Delta z}{D_S^*} \quad (8.14)$$

is most commonly selected to be identical with the restoring time scale (8.8). In case both fluxes, as given by (8.5) and (8.13), are applied in ocean models, we refer to *restoring boundary conditions*.

Analogously, atmosphere models require a condition for the lower boundary. Above water, it usually reads

$$F_W^{O \to A}(q) = D_W^* \left(q - q^* \right), \quad (8.15)$$

where q^* is a prescribed specific humidity. For land surfaces, simple hydrological models (*bucket models*) are commonly used. The coupling of atmosphere and ocean models requires that the salt fluxes in (8.13) are consistent with the water fluxes in (8.15). This is approximately accounted for by dividing the salt fluxes by a constant conversion factor ρS_0, S_0 is a reference salinity:

$$p - e = \frac{1}{\rho S_0} F_S^{O \to A} \quad (8.16)$$

and $p - e$ is the net water balance in m/s (or mm/yr). The distribution of $p - e$ is shown in Figs. 8.5 and 8.6. The conversion to energy fluxes is done according to

$$E = \rho L e, \quad P = \rho L p, \quad (8.17)$$

with $L = 2.5 \cdot 10^6$ J kg^{-1} for the specific latent heat of water.

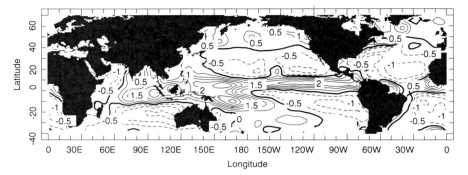

Fig. 8.5 Distribution of water fluxes $p-e$ in meter per year, at a contour interval of 0.5 m per year. The map was compiled on the internet (http://ingrid.ldeo.columbia.edu), where numerous data sets are available (here we used the Oberhuber data set).

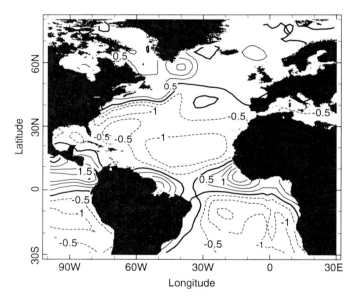

Fig. 8.6 Water fluxes $p-e$ in the Atlantic in meter per year. The map was compiled on the internet (http://ingrid.ldeo.columbia.edu), where numerous data sets are available (here we used the Oberhuber data set).

8.4 Momentum Fluxes

Any wind stress τ on the ocean surface is due to a momentum flux between ocean and atmosphere. It is a function of the horizontal wind speed $v_h = \sqrt{u^2 + v^2}$. For dimensional reasons of the units, the following parameterisation is used,

$$\tau = c_D \rho v_h^2 \,, \tag{8.18}$$

8.5 Mixed Boundary Conditions

where ρ is the density of air and c_D is a dimensionless transfer coefficient for momentum. Based on wind tunnel experiments, one may select for atmosphere–ocean fluxes

$$c_D = \begin{cases} 1.1 \cdot 10^{-3} & 0 < |u| < 6\,\text{m}\,\text{s}^{-1} \\ 0.61 \cdot 10^{-3} + 6.3 \cdot 10^{-5}\,\text{s}\,\text{m}^{-1} \cdot |u| & 6\,\text{m}\,\text{s}^{-1} < |u| < 22\,\text{m}\,\text{s}^{-1}, \end{cases} \quad (8.19)$$

but many other parameterisations have been proposed to account for different characteristics of the interface. In the models, vertical momentum fluxes are implemented as forces acting on the uppermost layer of the ocean model (or the lowest layer in the atmosphere model).

8.5 Mixed Boundary Conditions

Restoring boundary conditions as given by (8.5) are useful when equilibria of ocean models are sought for which the surface temperatures and salinities should be in good agreement with the data. Analogously, they are used in atmosphere-only models when sea surface temperatures are prescribed. For heat fluxes it may plausibly be argued that fluxes are proportional to the deviations. In fact, this is a discretized formulation of the heat flux according to Fick's first law. In physical terms, this means that for example a warm anomaly of the surface temperature in the ocean leads to an increased heat flux from the ocean to the atmosphere and hence causes a cooling of the ocean tending to restore the previous equilibrium.

However, the same argument cannot be used for water fluxes. A locally increased salinity at the surface of the ocean, for example induced by an oceanic eddy, does not lead to increased precipitation (Fig. 8.7). Such anomalies are therefore not eliminated on a typical time scale and have a much longer life time. Hence, (8.13) lacks a physical justification.

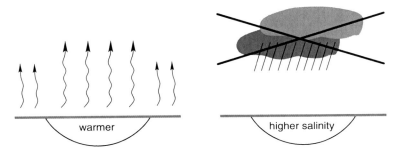

Fig. 8.7 Schematic illustration of temperature and salinity anomalies at the surface ocean and the different responses of heat and water fluxes. A warm SST anomaly causes an increased ocean-to-atmosphere heat flux which removes the anomaly. On the other hand, a SSS anomaly does not influence the amount of precipitation in the atmosphere.

In order to account for this fact in ocean models that have reached equilibrium (after several 1,000 years for T and S), (8.13) is replaced by a constant flux

$$\hat{F}_S^{O \to A} = D_S^* \left(S_\infty - S^* \right), \tag{8.20}$$

where S_∞ is the salinity attained in equilibrium. According to (8.20), $\hat{F}_S^{O \to A}$ is not time-dependent. The combination of the two boundary conditions (8.5) and (8.20) is denoted *mixed boundary conditions*. In principle, they represent a first approximation to the different nature of feedback processes associated with heat and water fluxes.

For mixed boundary conditions, the salinity, and hence the density at the surface ocean can deviate arbitrarily from a fixed prescribed distribution S^* without water fluxes to react and to counteract the emerging changes. This implies that salinity anomalies could permanently alter the structure of the circulation. This concept – first proposed by Stommel (1961) – was used by Bryan (1986) in a three-dimensional ocean model. The surprising result was the detection of *multiple equilibria*: For different boundary conditions, qualitatively different ocean circulations were simulated. This is further discussed in Chap. 9.

8.6 Coupled Models

The biggest challenge in climate modelling is the construction of consistent coupled models that incorporate and quantitatively simulate the components ocean, atmosphere, cryosphere, land surface, biosphere as well as the physical–biogeochemical interactions. Over the years, large progress in the coupling has been achieved as illustrated in Fig. 1.11. A particular difficulty is to simulate climatologies of the ocean as well as the atmosphere, that agree well with the observations. For a long period, the fact that the ocean and the atmosphere required different fluxes in order to reach equilibrium, was a major obstacle in modelling. This implies, that at the time of coupling, the two model components cannot be driven by the same fluxes. This inevitably leads to drift of the two components and possibly to completely unrealistic states.

Ideally, the coupling follows the scheme presented in Fig. 8.8 without flux corrections. In climate models of earlier generations, this often led to climate drift. A stable state agreeing with the climatology could not be attained. This is especially difficult, when the model state is in a range where several equilibria are possible.

Such a climate drift simulated with a coupled model of reduced complexity is shown in Fig. 8.9. The ocean component is first brought to equilibrium for 4,000 years under *restoring boundary conditions*. Subsequently, a simple energy balance model is coupled to the ocean model. From this point on, more degrees of freedom are available, therefore, T and S may also change.

As mentioned earlier, the prevention of drift is based on an unphysical crutch: so-called *flux corrections*. Although the latest generation of coupled models no

8.6 Coupled Models

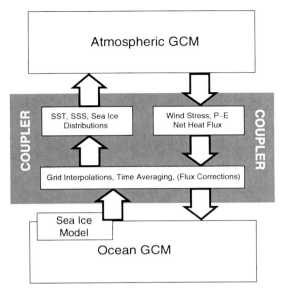

Fig. 8.8 Scheme of the coupling of atmosphere and ocean model components. Figure from Trenberth (1992).

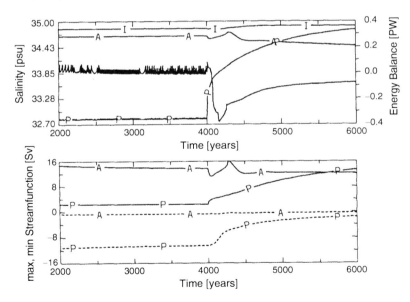

Fig. 8.9 Climate drift in a coupled model of reduced complexity, the Bern 2.5d model, initiating at the time of coupling ($t = 4,000$ years). The salinity of the Pacific (P) is increasing, the energy balance is perturbed and an approach to a different equilibrium state is simulated (**upper panel**). This leads to a change in the meridional circulation (**lower panel**), apparent in the drift of the stream functions. Figure from Stocker et al. (1992).

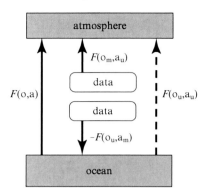

Fig. 8.10 Schematic depiction of the different fluxes with a flux correction in coupled climate models; "data" denotes measured climatologies. Adapted from Sausen et al. (1988).

longer requires such flux corrections, they shall be discussed more thoroughly, also with respect to their order of magnitude.

Flux corrections for heat, water and momentum fluxes are implemented as constant artificial sources and sinks at the boundaries of the individual model components. In doing so, the different model components are not fully coupled, but are rather linked via their deviations from an independently maintained equilibrium state. This is also referred to as *anomaly coupling*.

Flux corrections may be described following Sausen et al. (1988) and Egger (1997). $F(o, a)$ denotes the heat flux from the ocean (o) to the atmosphere (a), as it results from a fully coupled model, hence it is computed based on the variables T_O and T_A in (8.1). $F(o_u, a_u)$ denotes the heat flux that the *uncoupled* model requires in which $T_{O,u}$ and $T_{A,u}$ are used. In contrast, $F(o_u, a_m)$ denotes the heat flux based on fixed observational data of the atmosphere based on *measurements* (a_m) and values of the *uncoupled* ocean model (o_u) and $F(o_m, a_u)$ analogously for the *uncoupled* atmosphere model. This is illustrated in Fig. 8.10.

Instead of driving the ocean model with the fully coupled fluxes $F(o, a)$, $F(o, a)$ is replaced by

$$\tilde{F}_o(o, a) = F(o, a) + \underbrace{\{F(o_u, a_m) - F(o_u, a_u)\}}_{FO}, \tag{8.21}$$

where FO is the flux correction for the ocean model. $FO = 0$, in case the variables from the uncoupled atmosphere model completely agree with the measured quantities, i.e., $a_u = a_m$. Analogously, for the atmosphere model we write

$$\tilde{F}_a(o, a) = F(o, a) + \underbrace{\{F(o_m, a_u) - F(o_u, a_u)\}}_{FA}, \tag{8.22}$$

where FA is the flux correction for the atmosphere model. The difference $\tilde{F}_o - \tilde{F}_a$ is the artificial net source of heat, induced by the deviations of the modeled fluxes in the uncoupled model from the measured fluxes. The corrections in (8.21) and (8.22) may reach the same order of magnitude as the fluxes themselves. For the heat flux, this is shown in Fig. 8.11, for the flux of water in Fig. 8.12.

8.6 Coupled Models

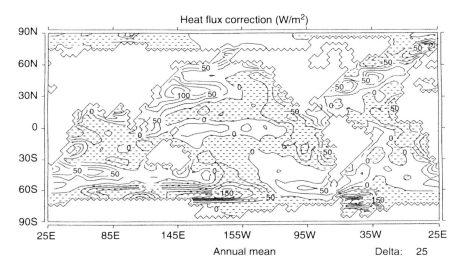

Fig. 8.11 Correction of the heat flux. Particularly in areas with strong oceanic currents (Gulf stream and Kuroshio), as well as in areas of deep water formation (Norwegian and Weddell Seas), very large fluxes result. Figure from Schiller et al. (1997).

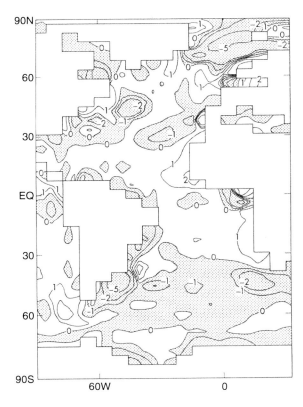

Fig. 8.12 Correction of the water flux in meter per year. Particularly in areas with strong oceanic currents (Gulf stream and Brazil Current), as well as in areas of deep water formation (Norwegian Sea), very large fluxes result. Figure from Manabe and Stouffer (1988).

By means of simple models and other considerations, it can be demonstrated, that for relatively small changes (e.g., the simulation of the next 50 year's climate) flux corrections do not yield fundamentally different results compared to the ones without flux correction (Egger 1997). However, one has to be generally cautious when interpreting such models.

As mentioned above, most of current climate models no longer employ flux corrections. This is an evidence for the progress in the understanding of processes in the climate system components (ocean, atmosphere, land, sea ice, snow, vegetation, etc.) and their representation in coupled models. Improved parameterisations, and to some extent a higher model grid resolution, have contributed to this progress.

However, occasionally coupled models still use other forms of flux correction, e.g., an imposed additional freshwater flux from the Atlantic to the Pacific in order to enhance deep water formation in the North Atlantic and improve the global ocean circulation (e.g., Zaucker et al. 1994; Renssen et al. 2005; Ritz et al. 2011).

Chapter 9
Multiple Equilibria in the Climate System

9.1 Abrupt Climate Change Recorded in Polar Ice Cores

The most detailed information about past climate states of the last 800,000 years can be retrieved from polar ice cores (Jouzel et al. 2007). One example for the last 90,000 years is presented in Fig. 9.1. The Holocene, the present interglacial, has started after the abrupt end of the last glacial period, 11,650 years ago. The transition from the last ice age to the Holocene, called Termination I, started about 20,000 years ago. An increase in the concentrations of particular isotopes could be detected in Antarctic ice cores. Stable isotopes of the water molecule are a measure for the local temperature. The temperature indicators also show that the climate changed in an abrupt way 25 times in Greenland during the last glacial period. These abrupt warming events, numbered in Fig. 9.1, are now referred to as *Dansgaard–Oeschger events (D/O events)* in remembrance of the research of the two pioneers in ice core science *Willy Dansgaard* (1922–2011) and *Hans Oeschger* (1927–1998) from the University of Copenhagen and the University of Bern.

These D/O events all show an abrupt warming of the northern hemisphere within one decade and a subsequent continuous cooling over about 1,000–3,000 years. Interestingly, the isotope maxima and minima during the glacial periods are all at the same level. Already in 1984, Hans Oeschger proposed that the climate system may have operated similar to a physical flip-flop and that the ocean circulation in the Atlantic Ocean is likely to be responsible for these climate jumps (Oeschger et al. 1984). Flip-flop systems are characterized by several stable equilibria. The Lorenz–Saltzman model (Sect. 7.2) is a classical example.

When Frank Bryan (1986) demonstrated using a three-dimensional ocean circulation model that several states of the thermohaline circulation can be realized, *Wally Broecker* synthesized the results from different climate archives and argued that rapid oscillations of the "Atlantic heat pump" (the thermohaline circulation) are responsible for the abrupt climate changes found in the Greenland ice cores, in tree rings, in sea and lake sediments, stalagmites, and in numerous other paleoclimatic archives (Broecker et al. 1985; Broecker and Denton 1989). Some relevant sources of related research on abrupt climate change are Alley et al. (2003), Barker et al. (2009), Blunier and Brook (2001), Blunier et al. (1998), Broecker (1997),

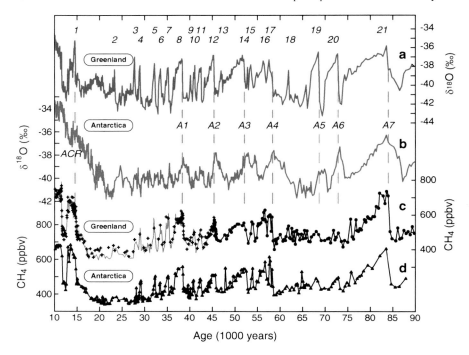

Fig. 9.1 Climate history of the last 90,000 years recorded in ice cores from Greenland and Antarctica. (**a**) Oxygen–isotope ratio ^{18}O (in per mille deviation from a predefined standard) in the GRIP ice core from Greenland; (**b**) ^{18}O in the Byrd core from Antarctica; (**c**) Methane concentration in the GRIP core; (**d**) Methane concentration in the Byrd core. In the Greenland ice core, 21 Dansgaard/Oeschger events are recorded. The longest D/O events exhibit a corresponding warm event in the Antarctic core; labeled A1 to A7. All of the D/O events are marked by abrupt peaks in the methane, enabling a synchronization of the time scales of Greenland and Antarctic ice cores. Figure from Blunier and Brook (2001).

Broecker and Denton (1989), Broecker et al. (1985), Clark et al. (2002), Dansgaard et al. (1984), Dansgaard et al. (1993), EPICA Community Members (2006), Huber et al. (2006), Knutti et al. (2004), Manabe and Stouffer (1988), Manabe and Stouffer (1994), Oeschger et al. (1984), Rahmstorf (2002), Stocker (1998), Stocker (2000), Stocker (2003), Stocker and Johnsen (2003), Stocker and Marchal (2000), Stocker and Wright (1991), Stocker et al. (1992).

9.2 The Bipolar Seesaw

Evidence from many climate archives supports the hypothesis that the ocean is primarily responsible for these abrupt changes. A sudden shut-down of the North Atlantic deep water formation causes a reduction of the meridional heat flux and therefore an abrupt cooling in the North Atlantic region. An active meridional

9.2 The Bipolar Seesaw

Fig. 9.2 Bipolar seesaw coupled with a southern heat reservoir to form the thermal bipolar seesaw. Figure from Stocker and Johnsen (2003).

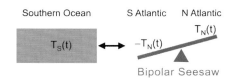

current draws heat from the Southern Atlantic. A shut-down of the heat pump will consequently cause a warming of the Southern Atlantic and should be noticeable in distinct teleconnections. This has led to the formulation of the so-called "Bipolar Seesaw" as a paradigm for the interaction of the northern and southern hemisphere during abrupt climate transitions (Broecker 1998; Stocker 1998). The bipolar seesaw is shown in Fig. 9.2 (right part) and suggests that an abrupt warming in the north leads to an abrupt cooling of the Southern Atlantic and vice-versa. This hypothesis makes distinct predictions that can be tested in climate archives.

A slightly more elaborate concept is the thermal bipolar seesaw proposed by Stocker and Johnsen (2003). It results from coupling a large heat reservoir to the southern end of the seesaw and leads to a fundamentally different temporal response of the Southern Ocean to abrupt temperature changes in the north. An abrupt cooling in the South Atlantic (i.e., abrupt warming in the North Atlantic) induces a slow continuous cooling in the whole Southern Ocean. In this simple manner, the very different characteristics of temperature signals extracted from ice cores of Greenland and Antarctica and shown in Fig. 9.1 can be explained.

The thermal bipolar seesaw is formulated as an energy balance for the Southern Ocean temperature:

$$\frac{dT_S(t)}{dt} = \frac{1}{\tau}\bigl(-T_N(t) - T_S(t)\bigr), \tag{9.1}$$

where T_S is the temperature anomaly of the Southern Ocean and T_N may represent the temperature anomaly of Greenland. With this, $-T_N$ is the temperature anomaly of the South Atlantic adjacent to the Southern Ocean assuming the instantaneous seesaw. τ is a characteristic time scale for the heat equilibration in the Southern Ocean. If $T_N(t)$ is given, the temporal evolution of $T_S(t)$ can be determined by a Laplace transformation of (9.1):

$$T_S(t) = -\frac{1}{\tau} \int_0^t T_N(t-t')\,e^{-t'/\tau}\,dt' + T_S(0)\,e^{-t/\tau}. \tag{9.2}$$

Hence, T_S is completely determined by the temporal evolution of T_N and reflects the northern temperature with a "damped memory". Let us consider this simple model in order to explain the different temporal evolution of the temperatures in Greenland and the Antarctica. By tuning the only free parameter τ we aim at producing the largest possible correlation between the modeled T_S based on (9.2) with the known temperature from the ice core T_N as input and the measured T_S derived from the Antarctic ice core. For $\tau \approx 1{,}100$ years a maximum correlation of

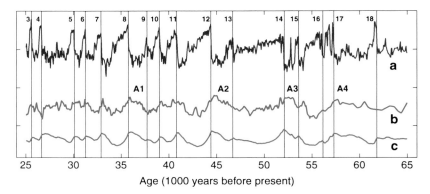

Fig. 9.3 High-pass filtered time series of the temperatures in Greenland (**a**) and Antarctica (**b**) derived from ice cores. (**c**) is the simulated temperature according to (9.2) with input (**a**). The abrupt Dansgaard–Oeschger events of the north hence become manifest in the local isotope maxima in Antarctica (A1, A2, ...). Figure from Stocker and Johnsen (2003).

0.77 is achieved. This allows us to predict the Antarctic temperature based on the temperature of Greenland in a surprisingly accurate way.

Although this simple concept explains a surprisingly large part of the variability, the required long time scale τ of over 1,000 years seems incompatible with the results from Ocean General Circulation Models (OGCMs), simulating only around 100 to 200 years as a typical exchange duration for the Southern Ocean.

There is another interesting consequence of the bipolar seesaw which follows from (9.2). Consider a very simple case of a northern temperature signal that has the shape of a periodic step function:

$$T_N(t) = \begin{cases} -\dfrac{1}{2}\Delta T & \text{for } (2i)\,t_0 < t < (2i+1)\,t_0 \\ +\dfrac{1}{2}\Delta T & \text{for } (2i+1)\,t_0 < t < 2(i+1)\,t_0 \end{cases} \quad i = 0,1,2,\ldots, \tag{9.3}$$

where ΔT is the temperature amplitude of abrupt changes in the north. In this case, we can determine $T_S(t)$ easily using (9.2). Assuming $T_S(0) = 0$ we get in the first interval $0 \le t < t_0$:

$$T_S(t) = \frac{1}{2}\Delta T\left(1 - e^{-t/\tau}\right). \tag{9.4}$$

Values for $T_S(t)$ in later intervals are calculated similarly. From the Taylor series expansion of this function about $t = 0$ truncated to first order, we obtain

$$\begin{aligned} T_S(t) &\approx T_S(0) + \left.\frac{dT_S}{dt}\right|_{t=0} t \\ &\approx \frac{\Delta T}{2\,\tau}\,t\,, \end{aligned}$$

9.2 The Bipolar Seesaw

which is a good approximation for $t \ll \tau$. We find a remarkable linear dependence of the maximum southern warming on the duration t_0 of the northern cooling,

$$T_S(t_0) \approx \frac{\Delta T}{2\tau} t_0 . \tag{9.5}$$

The longer the cooling in the northern Atlantic lasts due to the cessation of the meridional overturning circulation, the larger the warming will be in Antarctica. The warming further depends on the overall cooling, ΔT, in the north.

This linear relationship could be confirmed using the most recent information from the EPICA ice core from Dronning Maud Land (Antarctica). This ice core was drilled in a location geographically relatively close the Southern Atlantic Ocean where one would expect the largest influence of the bipolar seesaw. The duration of the stadials prior to the Dansgaard–Oeschger events was determined from the temperature reconstructions of the Greenland ice core from North GRIP; the amplitude of the warming in the south was obtained from the isotopic measurements on the EPICA ice core from Dronning Maud Land (EPICA Community Members 2006).

Figure 9.4 shows this impressive linear relationship for Marine Isotope Stage 3 during the last ice age and provides therefore the most convincing and independent evidence that much of the variability during an ice age can be captured by the very simple concept of the bipolar seesaw. It is remarkable that such a strong connection of the climatic behaviour on millennial time scales operates across the hemispheres.

More recent paleoclimatic reconstructions suggest that this mechanism also operated during the last Termination, i.e. the transition from the last ice age to the Holocene, a time period which was punctuated by large and abrupt climate changes such as the Bølling/Allerød warming and the Younger Dryas cooling in the North Atlantic region, and the Antarctic Cold Reversal in the south (Barker et al. 2009), as hypothesized earlier (Stocker 2003).

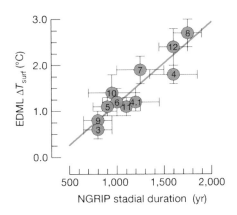

Fig. 9.4 Linear correlation between the duration of the cold stadials preceding the Dansgaard–Oeschger events in Greenland and the temperature amplitudes of the warmings in Antarctica. The numbers indicate the D/O events in Fig. 9.3. Figure from EPICA Community Members (2006).

9.3 Multiple Equilibria in a Simple Atmosphere Model

Geological evidence suggests that the Earth has gone through several phases of almost complete glaciation ("Snowball Earth" hypothesis, Hoffman and Li (2009)). How could this happen, given a roughly constant solar irradiation?

The energy balance model presented in Sect. 2.2 already yields a possible answer in case the ice-albedo feedback is accounted for (Sect. 2.4.1). Considering the equilibria of the energy balance (2.1) and parameterizing the albedo according to (2.22) but in a mathematically differentiable form, as illustrated in Fig. 2.12, an energy balance equation results that is non-linear in T:

$$\left(1 - \left(0.575 - 0.275 \tanh\left(0.033\,\text{K}^{-1}\,(T - 252.5\,\text{K})\right)\right)\right)\frac{S_0}{4} = \varepsilon\,\sigma\,T^4. \quad (9.6)$$

The left- and right-hand sides of (9.6) are shown in Fig. 9.5 for the two cases of a solar constant of $S_0 = 1367\,\text{W}\,\text{m}^{-2}$ and one which is reduced by 15%, $S_0^* = 0.85\,S_0$. For today's value of the solar constant (S_0) three equilibria exist, of which two are stable as indicated by filled circles. They represent a "warm" and a "cold" climate state. In the case of a 15% weaker solar constant (S_0^*, *faint young Sun*), only a single stable equilibrium exists corresponding to a very cold climate. Likewise, Fig. 9.5 reveals that the structure of the solution strongly depends on the specific form of the ice-albedo feedback parameterisation. For example, if an albedo parameterisation with a flatter slope were chosen, the two stable equilibria would shift towards the unstable one and finally merge into a single stable equilibrium.

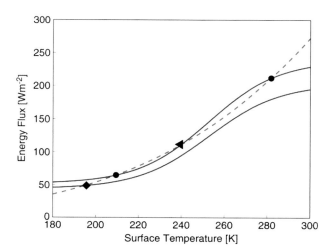

Fig. 9.5 Right-hand (*dashed*) and left-hand sides of (9.6) for S_0 (*upper curve*) and $0.85\,S_0$. The temperature dependence of the albedo produces several equilibrium solutions: stable states (*filled circles and diamond*) and an unstable equilibrium state (*triangle*).

9.4 Multiple Equilibria in a Simple Ocean Model

The question remains, whether multiple equilibria also exist in more complex climate models, e.g. in a coupled atmosphere–ocean model. This is discussed in Sect. 9.5.

9.4 Multiple Equilibria in a Simple Ocean Model

The deep circulation in the Atlantic is associated with a large heat transport that considerably affects climate in the North Atlantic region (conveyor belt). This heat transport is responsible for a comparatively mild climate. Already at the beginning of the twentieth century geologists assumed that the change in the ocean circulation may be responsible for part of the climate variability. In 1961, Henry Stommel presented a conceptual model that is able to reproduce such changes since it contains several equilibria (Stommel 1961). This model is presented in its simplified form following Marotzke (2000). The reason for the existence of multiple equilibria is linked to the fact that heat and water fluxes respond differently to anomalies. Mixed boundary conditions account for this phenomenon. Different relaxation times in (8.8) and (8.14) can also lead to several equilibria.

In this model the ocean is strongly simplified and consists of two boxes: one for latitudes where evaporation dominates (a positive water flux P) and one for the high latitudes where precipitation dominates (Fig. 9.6). T_i and S_i represent the temperatures and salinities of the two boxes, respectively. A fixed temperature difference ΔT between the boxes is assumed. It is maintained by heat fluxes between the atmosphere and the ocean. Between high and low latitudes a water transport q operates and is driven by the density difference according to

$$q = k(\rho_2 - \rho_1) = k \rho_0 \big(\alpha (T_2 - T_1) + \beta (S_2 - S_1)\big) \tag{9.7}$$

in which (8.11) with $\alpha < 0$ and $\beta > 0$ but with $\gamma = 0$ was used. The balance of the salinity in the two boxes is

$$\frac{dS_1}{dt} = |q|(S_2 - S_1) + P, \qquad \frac{dS_2}{dt} = |q|(S_1 - S_2) - P, \tag{9.8}$$

where $P > 0$ denotes the net water flux. In (9.8) the absolute value of q appears, because for the exchange the direction of the current is irrelevant. Stationary

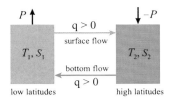

Fig. 9.6 2-Box model of the thermohaline circulation. Figure after Stommel (1961) and Marotzke (2000).

solutions for (9.8) can only be found if

$$\Delta S = S_2 - S_1 = \begin{cases} -\dfrac{\alpha \Delta T}{2\beta} \pm \sqrt{\left(\dfrac{\alpha \Delta T}{2\beta}\right)^2 - \dfrac{P}{\rho_0 k \beta}} & q > 0 \\ -\dfrac{\alpha \Delta T}{2\beta} - \sqrt{\left(\dfrac{\alpha \Delta T}{2\beta}\right)^2 + \dfrac{P}{\rho_0 k \beta}} & q < 0 , \end{cases} \quad (9.9)$$

where $\Delta T = T_2 - T_1 < 0$. For the *direct circulation*, $q > 0$ and hence $\rho_2 > \rho_1$, two solutions are possible: one with a smaller contrast in salinity $\Delta S < 0$ and one with a large negative ΔS. For an even smaller ΔS an *indirect circulation* exists, $q < 0$ and $\rho_2 < \rho_1$. We put

$$\delta = -\frac{\beta \Delta S}{\alpha \Delta T} , \quad E = \frac{\beta P}{\rho_0 k (\alpha \Delta T)^2} , \quad (9.10)$$

and obtain from (9.9)

$$\delta = \begin{cases} \dfrac{1}{2} \pm \sqrt{\dfrac{1}{4} - E} & q > 0 \\ \dfrac{1}{2} + \sqrt{\dfrac{1}{4} + E} & q < 0 . \end{cases} \quad (9.11)$$

The transport q is given by

$$q = k \rho_0 \alpha \Delta T (1 - \delta) . \quad (9.12)$$

For $\delta > 1$ the circulation is *indirect*, i.e., water sinks where it is warmer. In order to attain a sufficiently high density that permits a sinking, the salinity must be accordingly high. For $\delta < 1$ two solutions result, of which one is unstable (Fig. 9.7). For the *direct* circulation (water sinks where it is colder) $q > 0$. In case P increases, E and δ increase as well. But this leads to a decrease of q. An amplified hydrological cycle slows down the thermohaline circulation.

For $0.5 < \delta < 1$ and hence $0 < q < \frac{1}{2} k \rho_0 \alpha \Delta T$ the circulation is unstable. The model shows a threshold for q, below which the thermohaline circulation does not exist. It must be noted, that in this simple model the meridional temperature contrast directly determines this threshold.

The existence of multiple equilibria of the thermohaline circulation can be qualitatively understood by considering the heat and water transport as schematically illustrated in Fig. 9.8. In Fig. 9.8a the circulation is direct. Under the typical depth-profiles of T and S in the ocean (with respectively high values at the surface) the ocean transports heat and salt northwards. The cycle of the fluxes is closed by an excess of heat in the equatorial region and a cooling in the north, and by the

9.5 Multiple Equilibria in Coupled Models

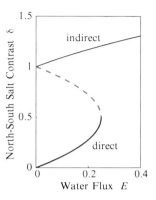

Fig. 9.7 Multiple equilibria (unstable = *dashed*) of the thermohaline circulation for different values of the water flux in the Stommel box model.

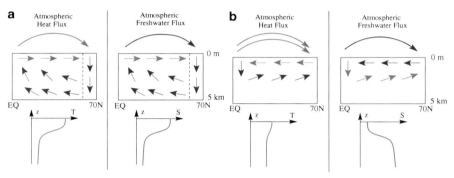

Fig. 9.8 Schematic depiction of the thermohaline circulation and the meridional heat and water fluxes. (**a**) direct circulation: water sinks where it is cold; (**b**) indirect circulation: water sinks in warm areas. Arrows of the water circulation are color coded: *red* (warm), *blue* (cold/fresh), *green* (salty).

atmospheric water transport. However, the same water transport can also result from an opposite circulation as shown in (b) in case the vertical gradient of S changes sign.

Hence, significant relocations of salt masses in the ocean are necessary in order to provoke basin-scale changes in the oceanic circulation. In the context of mixed boundary conditions for ocean models the salinity at the surface may change in an unlimited way which is a precondition for attaining state (b) in Fig. 9.8. The question whether this bears any realism is addressed in the next section.

9.5 Multiple Equilibria in Coupled Models

Model simulations by Manabe and Stouffer (1988) revealed for the first time results from a coupled climate model, in which for present climate conditions, two different states were found. They primarily differed in their thermohaline circulation in the

Atlantic. One of the states had an active deep water formation in the North Atlantic, the other state showed a circulation similar to the one in the Pacific. Transitions can be triggered by short-term differences in the water balance in the North Atlantic. Similar results were also found with other coupled models.

Therefore, it is probable that the deep water circulation in the Atlantic sensitively responds to changes in the surface water balance. This is a plausible mechanism to explain the abrupt changes found in climate time series (e.g., Fig. 9.1). One hypothesis claims that during glacial periods the ice sheets located around the North Atlantic discharged large amounts of freshwater caused by advancing ice streams. This situation was reinforced towards the end of the last glacial period, when the melting of the northern hemispheric ice sheets led to a sea level rise of about 120 m. During that time, the last sequence of abrupt climate changes was observed.

The lessons from the past clearly raise the question whether limited stability of the climate system, observed in many paleoclimatic records, may also be an issue today when the increase of greenhouse gas concentrations represents a significant perturbation to the climate system. The anthropogenic warming in the atmosphere not only increases sea surface temperatures but also alters the freshwater balance in a profound way. First, the melting of Greenland, which is proceeding at rapid rates, delivers freshwater to the Atlantic Ocean. Second, a warmer climate intensifies the water cycle due to the increased amount of water vapour in the atmosphere and to the higher evaporation rates induced by higher temperatures. This leads to a stronger meridional transport of water in the atmosphere. All three processes (warming of the SST, melting of Greenland and more precipitation) tend to decrease the sea surface density in the North Atlantic and, in consequence, have the potential to reduce the formation of deep water in the North Atlantic.

The question remains, whether this has basin-scale implications with the possibility that the Atlantic meridional overturning circulation may weaken in the future. Whether a threshold will be exceeded and a complete shut-down of this circulation system follows, is the object of current research.

The Intergovernmental Panel on Climate Change has addressed this issue (IPCC 2007, Chap. 10). Figure 9.9 illustrates the change in the meridional overturning circulation of the Atlantic for the coming 200 years based on different coupled models. Large differences between models exist; some models are inconsistent with observational estimates of the Atlantic meridional overturning.

Nevertheless, a general weakening trend during the twenty-first century emerges. None of the models simulates an intensification or an abrupt shut-down under this scenario within the coming 100 to 200 years.

Models of reduced complexity show that a threshold of the circulation exists beyond which a complete shut-down of the current results without additional external inputs. Therefore, a transition to a second stable equilibrium occurs. This behaviour might also be observable in more complete models (three-dimensional coupled atmosphere–ocean models without flux corrections), according most recent simulations (Mikolajewicz et al. 2007). Multiple equilibria were also shown in a fully coupled AOGCM, although in an aquaplanet configuration (Ferreira et al. 2011).

9.5 Multiple Equilibria in Coupled Models

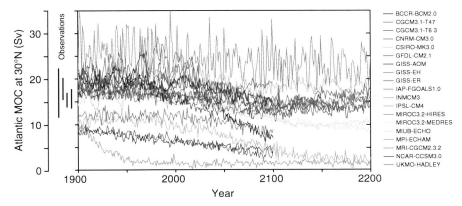

Fig. 9.9 Changes in the Atlantic meridional overturning circulation in 19 coupled climate models for the warming scenario A1B until year 2100, holding greenhouse gas concentrations constant thereafter. Many models show a weakening, but the uncertainties remain relatively high. Some models do not agree with observation-based estimates (*black vertical bars*). Figure from IPCC (2007), Chap. 10 (Fig. 10.15, p. 773).

Simulations with a simplified coupled model (Bern 2.5d model), consisting of a zonally averaged 3-basin ocean model and an energy balance model for the atmosphere, show that the threshold depends on several important quantities in the climate system, as well as the history of the perturbation. Figure 9.10 gives a summary of the results. For the respective simulations, simplified CO_2-scenarios were chosen: after an exponential growth at different rates, the CO_2 concentration was held constant. The evolution of the thermohaline circulation may be split into two cases. One exhibits a linear behaviour in which a temporarily strong reduction of the circulation is followed by a recovery over a few centuries. The reduction of the overturning circulation depends on the maximum value of the CO_2 increase and hence on the warming. In the second case, the circulation shuts down completely and does not recover. An irreversible transition to the second stable equilibrium has been realized.

It is interesting to notice that a reduction of the maximum concentration of CO_2 (from experiment 750 to 650) as well as a reduction of the rate of CO_2 increase (from experiment 750 to 750S) avoids the crossing of the critical threshold. Hence, the rate of future warming in the climate system plays a significant role. Depending on the rate and amount of warming, irreversible changes may result.

There is evidence that a similar behaviour can be produced by a more complex model. However, it must be considered that these models contain more degrees of freedom and hence respond to disturbances in a much more sophisticated way. The question, whether multiple equilibria can also occur in the models of highest resolution, remains unresolved.

In its latest assessment report, the IPCC draws a cautious conclusion regarding this problem in the Summary for Policymakers, IPCC (2007):

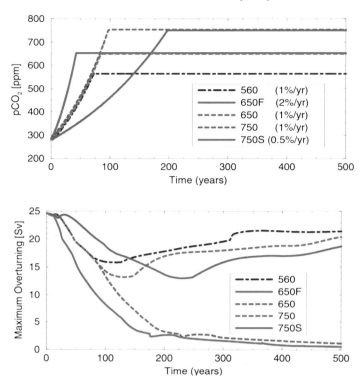

Fig. 9.10 Simulations with the Bern 2.5d model for the evolution of the meridional overturning circulation (MOC) in the Atlantic considering a warming scenario. The different simplified CO_2-scenarios (**upper panel**) consist of an exponential increase at different rates, leveling off at a given maximum value. The MOC reveals a bifurcation in its behaviour (**lower panel**): for small maximum values or slow rates of CO_2 increase, the threshold for a complete shut-down may be avoided. Figure from Stocker and Schmittner (1997).

> Based on current model simulations, it is *very likely* that the meridional overturning circulation (MOC) of the Atlantic Ocean will slow down during the twenty-first century. The multi-model average reduction by 2100 is 25% (range from zero to about 50%) for SRES emission scenario A1B. Temperatures in the Atlantic region are projected to increase despite such changes due to the much larger warming associated with projected increases of greenhouse gases. It is *very unlikely* that the MOC will undergo a large abrupt transition during the twenty-first century. Longer-term changes in the MOC cannot be assessed with confidence.

Recent research has focused on the question whether there may exist other components in the climate system which may exhibit instabilities or which are forced into new, quite different, equilibrium states. One example of intensive debate is the fate of the Greenland ice sheet. Recent observations confirm sustained mass losses for both Greenland and Antarctica (Velicogna 2009). Some model simulations suggest that there exist thresholds for warming in the area of the Greenland ice sheet and if crossed, this may lead to an irreversible melt-down of the ice sheet with a massive sea level rise of more than 6 m over the next several 100 years. However, paleoclimatic information suggests that during the last interglacial about 120,000 years ago, which was about 4°C warmer than today, the Greenland ice sheet was still present, although much smaller in extent.

Also, the Amazonian rainforest is supposed to respond to anthropogenic climate change both directly to the warming and, of course, due to direct deforestation. Some model simulations suggest that in this area a steppe-like vegetation cover may develop which then would feed back to the regional hydrological cycle and produce a new state of stable, but much drier regional climate. Even if the large-scale climate conditions were reset to pre-industrial values, the change in vegetation would not be reversible.

Permafrost in the boral areas of Siberia and North America is also a system which is increasingly investigated. Large amounts of methane are trapped in the permafrost. With the warming, permafrost is melting which could release methane from these areas. As a powerful greenhouse gas, this would enhance the greenhouse effect. However, due to natural sinks for methane in the atmosphere and the short lifetime of methane (about 10 years), such a perturbation would disappear rather rapidly.

There is therefore the general concern that anthropogenic perturbations may have already caused irreversible climate change. In this context, one often refers to "Tipping Points" in the climate system (Lenton et al. 2008), although this concept is difficult to quantify in the climate system. Obviously, predictability is extremely low, if not impossible, for such climate instabilities.

9.6 Concluding Remarks

The goal of these Lecture Notes was to provide some basic knowledge in climate modelling. In addition to theoretical concepts and recent results from climate research, we have framed the material in a sequence of simple problems which were solved numerically. This afforded the opportunity to introduce some basic numerical solution techniques and expose specific characteristics of those. A further goal was to give an introduction to a few fundamental concepts of the dynamics of the climate system. Surely, these notes could only provide an initial, very limited insight into this fascinating topic. Hopefully, it was made clear, that questions remain unresolved and that therefore many areas of activity are open for good ideas and creative model design.

Climate modelling is the only, however by far not perfect, method to make quantitative statements concerning past climate change. For predictions of future changes, climate modelling is the only scientific basis. An ongoing analysis of observed data and climate variables, as well as a more profound understanding of the fundamental processes guarantees a continuous improvement of these models. The scientific assessment of the impact of human activities on this planet, and to foresee dangerous developments to a certain degree, becomes an important duty of a responsible modern society.

Climate models also help us design and develop a strategy for sustainability. This is necessary, because a stable climate is a crucial resource for humanity, even though this is not yet widely acknowledged. The stable climate is also a prerequisite for continuous ecosystem services. Each modification of a resource implies a risk. In this sense, the changes observed to date, should be considered as both a reminder and starting point for resolute actions. These are required if society decides that future warming should be limited (Copenhagen Accord, 2009).

References

Alley, R.B., J. Marotzke, W.D. Nordhaus, J.T. Overpeck, D.M. Peteet, R.A. Pielke Jr., R.T. Pierrehumbert, P.B. Rhines, T.F. Stocker, L.D. Talley, and J.M. Wallace (2003): Abrupt climate change. *Science*, **299**, 2005–2010.

Barker, S., P. Diz, M.J. Vautravers, J. Pike, G. Knorr, I.R. Hall, and W.S. Broecker (2009): Interhemispheric Atlantic seesaw response during the last deglaciation. *Nature*, **457**, 1097–1102.

Blunier, T. and E.J. Brook (2001): Timing of millennial-scale climate change in Antarctica and Greenland during the last glacial period. *Science*, **291**, 109–112.

Blunier, T., J. Chappellaz, J. Schwander, A. Dällenbach, B. Stauffer, T.F. Stocker, D. Raynaud, J. Jouzel, H.B. Clausen, C.U. Hammer, and S.J. Johnsen (1998): Asynchrony of Antarctic and Greenland climate change during the last glacial period. *Nature*, **394**, 739–743.

Bony, S., R. Colman, V.M. Kattsov, R.P. Allan, C.S. Bretherton, J.L. Dufresne, A. Hall, S. Hallegatte, M.M. Holland, W. Ingram, D.A. Randall, B.J. Soden, G. Tselioudis, and M.J. Webb (2006): How well do we understand and evaluate climate change feedback processes? *J. Climate*, **19**, 3445–3482.

Broecker, W.S. (1987): The biggest chill. *Nat. Hist.*, **96**, 74–82.

Broecker, W.S. (1997): Thermohaline circulation, the Achilles heel of our climate system: will man-made CO_2 upset the current balance? *Science*, **278**, 1582–1588.

Broecker, W.S. (1998): Paleocean circulation during the last deglaciation: a bipolar seesaw? *Paleoceanography*, **13**, 119–121.

Broecker, W.S. and G.H. Denton (1989): The role of ocean-atmosphere reorganizations in glacial cycles. *Geochim. Cosmochim. Ac.*, **53**, 2465–2501.

Broecker, W.S., D.M. Peteet, and D. Rind (1985): Does the ocean-atmosphere system have more than one stable mode of operation? *Nature*, **315**, 21–26.

Bryan, F. (1986): High-latitude salinity effects and interhemispheric thermohaline circulations. *Nature*, **323**, 301–304.

Bryan, F. (1987): Parameter sensitivity of primitive equation ocean general circulation models. *J. Phys. Oceanogr.*, **17**, 970–985.

Bryan, K. and M.D. Cox (1967): A numerical investigation of the oceanic general circulation. *Tellus*, **19**, 54–80.

Budyko, M.I. (1969): The effect of solar radiation variations on the climate of the earth. *Tellus*, **21**, 611–619.

Cayan, D.R. (1992): Latent and Sensible Heat Flux Anomalies over the Northern Oceans: The Connection to Monthly Atmospheric Circulation. *J. Climate*, **5**, 354–370.

Clark, P.U., N.G. Pisias, T.F. Stocker, and A.J. Weaver (2002): The role of the thermohaline circulation in abrupt climate change. *Nature*, **415**, 863–869.

Colman, R. (2003): A comparison of climate feedbacks in general circulation models. *Clim. Dyn.*, **20**, 865–873.

Copenhagen Accord (2009): Document available at http://unfccc.int.

Courant, R., K. Friedrichs, and H. Lewy (1928): Über die partiellen Differenzengleichungen der mathematischen Physik. *Math. Ann.*, **100**, 32–74.

Dansgaard, W., S.J. Johnsen, H.B. Clausen, D. Dahl-Jensen, N. Gundestrup, C.U. Hammer, and H. Oeschger (1984): North Atlantic climatic oscillations revealed by deep Greenland ice cores. *In:* J.E. Hansen (Editor), *Climate Processes and Climate Sensitivity (Geophysical Monograph)*, pp. 288–298. American Geophysical Union.

Dansgaard, W., S.J. Johnsen, H.B. Clausen, D. Dahl-Jensen, N.S. Gundestrup, C.U. Hammer, C.S. Hvidberg, J.P. Steffensen, A.E. Sveinbjörnsdottir, J. Jouzel, and G. Bond (1993): Evidence for general instability of past climate from a 250-kyr ice-core record. *Nature*, **364**, 218–220.

Domingues, C.M., J.A. Church, N.J. White, P.J. Gleckler, S.E. Wijffels, P.M. Barker, and J.R. Dunn (2008): Improved estimates of upper-ocean warming and multi-decadal sea-level rise. *Nature*, **453**, 1090–1093.

Egger, J. (1997): Flux correction: Tests with a simple ocean-atmosphere model. *Clim. Dyn.*, **13**, 285–292.

EPICA Community Members (2006): One-to-one coupling of glacial climate variability in Greenland and Antarctica. *Nature*, **444**, 195–198.

Ferreira, D., J. Marshall, and B. Rose (2011): Climate determinism revisited: Multiple equilibria in a complex climate model. *J. Climate*, **24**, 992–1012.

Ganachaud, A. and C. Wunsch (2000): Improved estimates of global ocean circulation, heat transport and mixing from hydrographic data. *Nature*, **408**, 453–457.

Gill, A.E. (1982): *Atmosphere–Ocean Dynamics*. Academic Press, 662 pp.

Haltiner, G.J. and R.T. Williams (1980): *Numerical Prediction and Dynamic Meteorology*. Wiley, 2 ed., 496 pp.

Haney, R.L. (1971): Surface thermal boundary condition for ocean circulation models. *J. Phys. Oceanogr.*, **1**, 241–248.

Hartmann, D.L. (1994): *Global Physical Climatology*. Academic Press, 1 ed., 411 pp.

Hoffman, P.F. and Z.X. Li (2009): A palaeogeographic context for Neoproterozoic glaciation. *Palaeogeogr. Palaeocl.*, **277**, 158–172.

Holton, J.R. (2004): *An introduction to dynamic meteorology*. Academic Press, 511 pp.

Houghton, J. (2001): *The Physics of Atmospheres*. Cambridge University Press, 3 ed., 336 pp.

Houghton, J. (2009): *Global Warming: The Complete Briefing*. Cambridge University Press, 4 ed., 456 pp.

Huber, C., M. Leuenberger, R. Spahni, J. Flückiger, J. Schwander, T.F. Stocker, S. Johnsen, A. Landais, and J. Jouzel (2006): Isotope calibrated Greenland temperature record over Marine Isotope Stage 3 and its relation to CH4. *Earth Planet. Sc. Lett.*, **243**, 504–519.

IPCC (2001): *Climate Change 2001 - The Scientific Basis: Contribution of Working Group I to the Third Assessment Report of the IPCC*. Cambridge University Press, 892 pp.

IPCC (2007): *Climate Change 2007 - The Physical Science Basis: Working Group I Contribution to the Fourth Assessment Report of the IPCC*. Cambridge University Press, 1009 pp.

Jayne, S.R. and J. Marotzke (2002): The oceanic eddy heat transport. *J. Phys. Oceanogr.*, **32**, 3328–3345.

Joos, F., G.K. Plattner, T.F. Stocker, O. Marchal, and A. Schmittner (1999): Global warming and marine carbon cycle feedbacks on future atmospheric CO_2. *Science*, **284**, 464–467.

Jouzel, J., V. Masson-Delmotte, O. Cattani, G. Dreyfus, S. Falourd, G. Hoffmann, M. Minster, J. Nouet, J.-M. Barnola, J. Chappellaz, H. Fischer, J.C. Gallet, S. Johnsen, M. Leuenberger, L. Loulergue, D. Lüthi, H. Oerter, F. Parrenin, G. Raisbeck, D. Raynaud, A. Schilt, J. Schwander, E. Selmo, R. Souchez, R. Spahni, B. Stauffer, J.P. Steffensen, B. Stenni, T.F. Stocker, J.L. Tison, M. Werner, and E.W. Wolff (2007): Orbital and millennial Antarctic climate variability over the past 800,000 years. *Science*, **317**, 793–796.

Knutti, R. and G.C. Hegerl (2008): The equilibrium sensitivity of the EarthŠs temperature to radiation changes. *Nat. Geosci.*, **1**, 735–743.

Knutti, R., T.F. Stocker, F. Joos, and G.-K. Plattner (2002): Constraints on radiative forcing and future climate change from observations and climate model ensembles. *Nature*, **416**, 719–723.

References

Knutti, R., T.F. Stocker, F. Joos, and G.K. Plattner (2003): Probabilistic climate change projections using neural networks. *Clim. Dyn.*, **21**, 257–272.

Knutti, R., J. Flückiger, T.F. Stocker, and A. Timmermann (2004): Strong hemispheric coupling of glacial climate through freshwater discharge and ocean circulation. *Nature*, **430**, 851–856.

Kraus, E.B. and J.A. Businger (1994): *Atmosphere–Ocean Interaction*. Oxford University Press, 362 pp.

Krishnamurti, T.N. and L. Bounoua (1996): *An Introduction to Numerical Weather Prediction Techniques*. CRC Press, 1 ed., 293 pp.

Lenton, T.M., H. Held, E. Kriegler, J.W. Hall, W. Lucht, S. Rahmstorf, and H.J. Schellnhuber (2008): Tipping elements in the Earth's climate system. *P. Natl. Acad. Sci. USA*, **105**, 1786–1793.

Lorenz, E.N. (1963): Deterministic nonperiodic flow. *J. Atmos. Sci.*, **20**, 130–141.

Lorenz, E.N. (1979): Forced and free variations of weather and climate. *J. Atmos. Sci.*, **36**, 1367–1376.

Lorenz, E.N. (1984): Irregularity: A fundamental property of the atmosphere. *Tellus*, **36**, 98–110.

Lorenz, E.N. (1996): *The Essence of Chaos*. University of Washington Press, 227 pp.

Lozier, M.S. (2010): Deconstructing the conveyor belt. *Science*, **328**, 1507–1511.

Lüthi, D., M. Le Floch, B. Bereiter, T. Blunier, J.-M. Barnola, U. Siegenthaler, D. Raynaud, J. Jouzel, H. Fischer, K. Kawamura, and T.F. Stocker (2008): High-resolution carbon dioxide concentration record 650,000–800,000 years before present. *Nature*, **453**, 379–382.

Manabe, S. and K. Bryan (1969): Climate calculations with a combined ocean-atmosphere model. *J. Atmos. Sci.*, **26**, 786–789.

Manabe, S. and R.J. Stouffer (1988): Two stable equilibria of a coupled ocean-atmosphere model. *J. Climate*, **1**, 841–866.

Manabe, S. and R.J. Stouffer (1994): Multiple-century response of a coupled ocean-atmosphere model to an increase of atmospheric carbon dioxide. *J. Climate*, **7**, 5–23.

Marchal, O., T.F. Stocker, F. Joos, A. Indermühle, T. Blunier, and J. Tschumi (1999): Modelling the concentration of atmospheric CO_2 during the Younger Dryas climate event. *Clim. Dyn.*, **15**, 341–354.

Marotzke, J. (2000): Abrupt climate change and thermohaline circulation: Mechanisms and predictability. *P. Natl. Acad. Sci. USA*, **97**, 1347–1350.

McGuffie, K. and A. Henderson-Sellers (2005): *A Climate Modelling Primer*. Wiley, 3 ed., 296 pp.

Mikolajewicz, U., M. Gröger, E. Maier-Reimer, G. Schurgers, M. Vizcaíno, and A.M.E. Winguth (2007): Long-term effects of anthropogenic CO_2 emissions simulated with a complex earth system model. *Clim. Dyn.*, **28**, 599–633.

Müller, S.A., F. Joos, N.R. Edwards, and T.F. Stocker (2006): Water mass distribution and ventilation time scales in a cost-efficient, three-dimensional ocean model. *J. Climate*, **19**, 5479–5499.

Myhre, G., E.J. Highwood, K.P. Shine, and F. Stordal (1998): New estimates of radiative forcing due to well mixed greenhouse gases. *Geophys. Res. Lett.*, **25**, 2715–2718.

Nakicenovic, N., J. Alcamo, G. Davis, B. de Vries, J. Fenhann, S. Gaffin, K. Gregory, A. Grubler, T.Y. Jung, T. Kram, E.L. La Rovere, L. Michaelis, S. Mori, T. Morita, W. Pepper, H. Pitcher, L. Price, K. Riahi, A. Roehrl, H.-H. Rogner, A. Sankovski, M. Schlesinger, P. Shukla, S. Smith, R. Swart, S. van Rooijen, N. Victor, and Z. Dadi (2000): *Special Report on Emissions Scenarios: A Special Report of Working Group III of the Intergovernmental Panel on Climate Change*. Cambridge University Press, 598 pp.

North, G.R., D.A. Short, and J.G. Mengel (1983): Simple energy balance model resolving the seasons and the continents - Application to the astronomical theory of the ice ages. *J. Geophys. Res.*, **88**, 6576–6586.

Oeschger, H., J. Beer, U. Siegenthaler, B. Stauffer, W. Dansgaard, and C.C. Langway (1984): Late glacial climate history from ice cores. In: J. E. Hansen and T. Takahashi (Editors), *Climate Processes and Climate Sensitivity*, vol. 29 of *Geophysical Monograph Series*, pp. 299–306. American Geophysical Union.

Pedlosky, J. (1987): *Geophysical Fluid Dynamics*. Springer, 710 pp.

Peixoto, J.P. and A.H. Oort (1992): *Physics of Climate*. Springer, 520 pp.
Petit, J.R., J. Jouzel, D. Raynaud, N.I. Barkov, J.-M. Barnola, I. Basile, M. Bender, J. Chappellaz, M. Davis, G. Delaygue, M. Delmotte, V.M. Kotlyakov, M. Legrand, V.Y. Lipenkov, C. Lorius, L. PÉpin, C. Ritz, E. Saltzman, and M. Stievenard (1999): Climate and atmospheric history of the past 420,000 years from the Vostok ice core, Antarctica. *Nature*, **399**, 429–436.
Phillips, N.A. (1956): The general circulation of the atmosphere: a numerical experiment, QJ Roy. *Q. J. Roy. Meteor. Soc.*, **82**, 123–164.
Plattner, G.K., R. Knutti, F. Joos, T.F. Stocker, W. von Bloh, V. Brovkin, D. Cameron, E. Driesschaert, S. Dutkiewicz, M. Eby, N.R. Edwards, T. Fichefet, J.C. Hargreaves, C.D. Jones, M.F. Loutre, H.D. Matthews, A. Mouchet, S.A. Müller, S. Nawrath, A. Price, A. Sokolov, K.M. Strassmann, and A.J. Weaver (2008): Long-term climate commitments projected with climate-carbon cycle models. *J. Climate*, **21**, 2721–2751.
Press, W.H., B.P. Flannery, S.A. Teukolsky, and W.T. Vetterling (1992): *Numerical Recipes in FORTRAN 77: The Art of Scientific Computing*, vol. 1. Cambridge University Press, 2 ed., 992 pp.
Press, W.H., S.A. Teukolsky, W.T. Vetterling, and B.P. Flannery (1996): *Numerical Recipes in Fortran 90: The Art of Scientific Computing*, vol. 2. Cambridge University Press, 2 ed., 500 pp.
Press, W.H., S.A. Teukolsky, W.T. Vetterling, and B.P. Flannery (2007): *Numerical Recipes 3rd Edition: The Art of Scientific Computing*. Cambridge University Press, 3 ed., 1256 pp.
Rahmstorf, S. (2002): Ocean circulation and climate during the past 120,000 years. *Nature*, **419**, 207–214.
Renssen, H., H. Goosse, T. Fichefet, V. Brovkin, E. Driesschaert, and F. Wolk (2005): Simulating the Holocene climate evolution at northern high latitudes using a coupled atmosphere-sea ice-ocean-vegetation model. *Clim. Dyn.*, **24**, 23–43.
Richardson, L.F. (2007): *Weather Prediction by Numerical Process (Cambridge Mathematical Library)*. Cambridge University Press, 2 ed., 250 pp. Reprint of original 1922.
Ritz, S.P., T.F. Stocker, and S.A. Müller (2008): Modeling the effect of abrupt ocean circulation change on marine reservoir age. *Earth Planet. Sc. Lett.*, **268**, 202–211.
Ritz, S.P., T.F. Stocker, and F. Joos (2011): A coupled dynamical ocean-energy balance atmosphere model for paleoclimate studies. *J. Climate*, **24**, 349–375.
Robock, A. (2008): 20 reasons why geoengineering may be a bad idea. *Bull. At. Sci.*, **64**, 14–18.
Robock, A., L. Oman, and G.L. Stenchikov (2008): Regional climate responses to geoengineering with tropical and Arctic SO_2 injections. *J. Geophys. Res.*, **113**, D16101.
Ruddiman, W.F. (2007): *Earth's Climate: Past and Future*. W.H. Freeman, 2 ed., 388 pp.
Saha, S., S. Nadiga, C. Thiaw, J. Wang, W. Wang, Q. Zhang, H. M. Van den Dool, H.-L. Pan, S. Moorthi, D. Behringer, D. Stokes, M. Peña, S. Lord, G. White, W. Ebisuzaki, P. Peng, and P. Xie (2006): The NCEP climate forecast system. *J. Climate*, **19**, 3483–3517.
Saltzman, B. (1962): Finite amplitude free convection as an initial value problem: I. *J. Atmos. Sci.*, **19**, 329–341.
Saltzman, B. (2001): *Dynamical Paleoclimatology: Generalized Theory of Global Climate Change*. Academic Press, 354 pp.
Sausen, R., K. Barthel, and K. Hasselmann (1988): Coupled ocean-atmosphere models with flux correction. *Clim. Dyn.*, **2**, 145–163.
Schär, C., P.L. Vidale, D. Lüthi, C. Frei, C. Häberli, M.A. Liniger, and C. Appenzeller (2004): The role of increasing temperature variability in European summer heatwaves. *Nature*, **427**, 332–336.
Schiller, A., U. Mikolajewicz, and R. Voss (1997): The stability of the North Atlantic thermohaline circulation in a coupled ocean-atmosphere general circulation model. *Clim. Dyn.*, **13**, 325–347.
Schwarz, H.R. (2004): *Numerische Mathematik*. Teubner B.G. GmbH, 573 pp.
Sellers, W.D. (1969): A global climatic model based on the energy balance of the earth-atmosphere system. *J. Appl. Meteorol.*, **8**, 392–400.
Siedler, G., J. Church, and J. Gould (Editors) (2001): *Ocean Circulation and Climate: Observing and Modeling the Global Ocean*, vol. 77 of *International Geophysics Series*. Academic Press, 715 pp.

References

Siegenthaler, U. and F. Joos (1992): Use of a simple model for studying oceanic tracer distributions and the global carbon cycle. *Tellus B*, **44**, 186–207.

Siegenthaler, U., T.F. Stocker, E. Monnin, D. Lüthi, J. Schwander, B. Stauffer, D. Raynaud, J.-M. Barnola, H. Fischer, V. Masson-Delmotte, and J. Jouzel (2005): Stable carbon cycle-climate relationship during the late Pleistocene. *Science*, **310**, 1313–1317.

Smolarkiewicz, P.K. (1983): A simple positive definite advection scheme with small implicit diffusion. *Mon. Weather Rev.*, **111**, 479–486.

Soden, B.J. and I.M. Held (2006): An assessment of climate feedbacks in coupled ocean–atmosphere models. *J. Climate*, **19**, 3354–3360.

Soden, B.J., R.T. Wetherald, G.L. Stenchikov, and A. Robock (2002): Global cooling after the eruption of Mount Pinatubo: A test of climate feedback by water vapor. *Science*, **296**, 727–730.

Stocker, T.F. (1998): The seesaw effect. *Science*, **282**, 61–62.

Stocker, T.F. (2000): Past and future reorganizations in the climate system. *Quaternary Sci. Rev.*, **19**, 301–319.

Stocker, T.F. (2003): South dials north. *Nature*, **424**, 496–499.

Stocker, T.F. and S.J. Johnsen (2003): A minimum thermodynamic model for the bipolar seesaw. *Paleoceanography*, **18**, 1087.

Stocker, T.F. and O. Marchal (2000): Abrupt climate change in the computer: Is it real? *P. Natl. Acad. Sci. USA*, **97**, 1362–1365.

Stocker, T.F. and A. Schmittner (1997): Influence of CO_2 emission rates on the stability of the thermohaline circulation. *Nature*, **388**, 862–865.

Stocker, T.F. and D.G. Wright (1991): A zonally averaged ocean model for the thermohaline circulation. Part II: Interocean circulation in the Pacific-Atlantic basin system. *J. Phys. Oceanogr.*, **21**, 1725–1739.

Stocker, T.F., D.G. Wright, and L.A. Mysak (1992): A zonally averaged, coupled ocean-atmosphere model for paleoclimate studies. *J. Climate*, **5**, 773–797.

Stommel, H. (1948): The westward intensification of wind-driven ocean currents. *Trans. Amer. Geophys. Union*, **29**, 202–206.

Stommel, H. (1958): The abyssal circulation. *Deep Sea Res.*, **5**, 80–82.

Stommel, H. (1961): Thermohaline convection with two stable regimes of flow. *Tellus*, **13**, 224–230.

Stommel, H. and A.B. Arons (1960): On the abyssal circulation of the world ocean–I. Stationary planetary flow patterns on a sphere. *Deep Sea Res.*, **6**, 140–154.

Stroeve, J., M.M. Holland, W. Meier, T. Scambos, and M. Serreze (2007): Arctic sea ice decline: Faster than forecast. *Geophys. Res. Lett.*, **34**, L09501.

Trenberth, K.E. (Editor) (1992): *Climate System Modeling*. Cambridge University Press, 1 ed., 818 pp.

Trenberth, K.E., J.M. Caron, and D.P. Stepaniak (2001): The atmospheric energy budget and implications for surface fluxes and ocean heat transports. *Clim. Dyn.*, **17**, 259–276.

Trenberth, K.E., J.T. Fasullo, and J. Kiehl (2009): Earth's global energy budget. *B. Am. Meteorol. Soc.*, **90**, 311–323.

UNFCCC (1992): *United Nations Framework Convention on Climate Change*. United Nations. URL http://unfccc.int.

Uppala, S.M., P.W. Kållberg, A.J. Simmons, U. Andrae, V.D.C. Bechtold, M. Fiorino, J.K. Gibson, J. Haseler, A. Hernandez, G.A. Kelly, X. Li, K. Onogi, S. Saarinen, N. Sokka, R.P. Allan, E. Andersson, K. Arpe, M.A. Balmaseda, A.C.M. Beljaars, L. Van De Berg, J. Bidlot, N. Bormann, S. Caires, F. Chevallier, A. Dethof, M. Dragosavac, M. Fisher, M. Fuentes, S. Hagemann, E. Hólm, B.J. Hoskins, L. Isaksen, P.A.E.M. Janssen, R. Jenne, A.P. Mcnally, J.-F. Mahfouf, J.-J. Morcrette, N.A. Rayner, R.W. Saunders, P. Simon, A. Sterl, K.E. Trenberth, A. Untch, D. Vasiljevic, P. Viterbo, and J. Woollen (2005): The ERA-40 re-analysis. *Q. J. Roy. Meteor. Soc.*, **131**, 2961–3012.

Velicogna, I. (2009): Increasing rates of ice mass loss from the Greenland and Antarctic ice sheets revealed by GRACE. *Geophys. Res. Lett.*, **36**, L19503.

Washington, W.M. and C.L. Parkinson (2005): *Introduction to Three-dimensional Climate Modeling*. University Science Books, 2 ed., 368 pp.

Wild, M. (2000): Absorption of solar energy in cloudless and cloudy atmospheres over Germany and in GCMs. *Geophys. Res. Lett.*, **27**, 959–962.

Willebrand, J. (1993): Forcing the ocean by heat and freshwater fluxes. *In:* E. Raschke and D. Jacob (Editors), *Energy and Water Cycles in the Climate System*, pp. 215–233. Springer.

Winton, M. (2006): Surface albedo feedback estimates for the AR4 climate models. *J. Climate*, **19**, 359–365.

Woodruff, S.D., R.J. Slutz, R.L. Jenne, and P.M. Steurer (1987): A Comprehensive Ocean-Atmosphere Data Set. *B. Am. Meteorol. Soc.*, **68**, 1239–1250.

Zaucker, F., T.F. Stocker, and W.S. Broecker (1994): Atmospheric freshwater fluxes and their effect on the global thermohaline circulation. *J. Geophys. Res.*, **99**, 12443–12457.

Index

Page numbers in **bold** denote figures and tables.

CH_4, see Methane
CO_2, see Carbon dioxide
H_2O, see Water

A-grid, 106, **107**
  one-dimensional energy balance model, **108**
Acceleration
  centrifugal, 99
  Coriolis, 99
  free-fall, 100
  gravity, 80, 99
Acronyms, list of, xv
Advection, 13, 27, 53, 55
  momentum, 13
Advection equation
  solutions, 61–75
    analytical, 61–62
    numerical, 63–75
Advection-diffusion equation, 56–57
  numerical solution, 75–76
Advection-diffusion model, 26, 27
Advective derivative, see Material derivative
Advective flux density, 55
  examples, **56**
Aerosols, 3, **16**, 42
  sulphate, 15
AGCM, see Atmospheric general circulation model
Albedo (reflectivity), 34–36, 42–44, 46, 47, 50, 84, 156
  ice, 43
  Sellers' parameterisation, 44, 156
Amazonian rainforest, 163
AMIP, see Atmospheric modelling intercomparison project

Angular momentum (of an air parcel), 123, 125, **127**, 128
  advective transport, 125
  conservation, 123, 125
Anomaly coupling, 148
Antarctic Cold Reversal, 155
Antarctica, 7, 153, 155, **155**, 163
  Dronning Maud Land, 155
  ice core, 6, 21, **23**, 151, **152**, 153
  ice sheet, 21
  temperature, 154, **154**
Anthroposphere, 4
AOGCM, see Atmosphere/Ocean general circulation model
Apparent force, see Inertial force
Arakawa grid (A-, E-, C-), 106, **107**
  one-dimensional energy balance model, **108**
Archimedes' principle, 130
Arctic sea ice
  decrease, 16–17
  since 1900, **17**
Atmosphere, 3
Atmosphere–Ocean interactions, 137–150
  fluxes of water, heat and momentum, **138**
Atmosphere/Ocean general circulation model (AOGCM), 28, **31**, **50**
Atmospheric general circulation model (AGCM), **26**, 28, **29**, 32, **32**, 37, 85
Atmospheric modelling intercomparison project (AMIP), 25

Backward difference, see Numerical scheme, Euler backward difference
Baroclinic fluid, 120
Baroclinic instabilities, 88

Barotropic fluid, 120
Barotropic gyre, 86
Barotropic instabilities, 88
Barotropic ocean circulation, 120
Basis functions, 109, 110
Bern 2.5-dimensional model, ix, 8, **19**, 26, **28**, 147, 161, **162**
β-plane, *see also* f-plane, 99, **99**, 111
Biosphere, 4, 19, 146
 "breathing", 21
Bipolar seesaw, **153**, 152–155
 thermal, 153, **153**
Bjerknes, Vilhelm, 10, **10**, 11
Boundary
 -value problem, 91–96, 113–116
 condition, 79, 84, 92, 108–110, 113, 126, 127, 131, 137, 143, 146, 157, 159
  Cauchy, 92
  Dirichlet, 92, 93, 113, 115
  hydrological, 142–143
  mixed, 145–146
  Neumann, 92, 114, 115
  periodic, 105
  restoring, 143, 145, 146
  thermal, 137–142
 layer, 137
  Stommel, 113
Brazil current, 110
Broecker, Wally, viii, 85, 151
Brownian motion, 53, **54**
Brunt–Väisälä frequency, 126
Bryan, Frank, 151
Bucket model, 143
Buoyancy, 126, 130
"butterfly effect", 13
Bølling/Allerød warming, 155

C^4MIP, *see* Coupled climate-carbon cycle modelling intercomparison project
C-grid, 106, **107**
 one-dimensional energy balance model, **108**
Cape Hatteras, 121
Carbon cycle, 15, 19, 27, 28
Carbon dioxide (CO_2), 1, 4, 5, 7, 8, 15, 16, 18, 19, 21, 22, 25, 27, 28, 36, 41, 42, 49, 50, 58, 161
 doubling the concentration, 42, 46, 49
 emissions leading to stabilization, 19, **19**
 link to changes in temperature, 15
 scenarios, 161, **162**
Centered/central difference, *see* Numerical scheme, centered difference

Central difference, 40
Centrifugal acceleration, 99
Centrifugal force, 98–100
CFL criterion, *see* Courant–Friedrichs–Lewy criterion
Chaos theory, 13, 128
Climate archives, 151–153
 ice cores, 7, 151
 sediments, 7, 151
 speleothems (e.g. stalagmites), 7, 151
 tree rings, 7, **9**, 151
Climate change, 7, 21, 22, 27, 28, 84, **161, 162**
 abrupt, viii, 151–152, 160
 anthropogenic, 3, 8, 15, 22, 28, 160, 163
 current, 8
 future, vii, 1, 7, 15, 27
 past, vii, 8, **9, 16, 17, 23**, 27, **152, 154, 155**, 163
 summer temperatures over Europe, **18**
Climate drift, 146, **147**
Climate feedback, 41–51, 146, 163
 cloud, 43, 45–48
 ice-albedo, 43–45, 50, 156
 lapse rate, 47–48, 50
 parameter, 42–44
 Planck, 43, 48
 snow-albedo, 17
 soil moisture, 17
 summary and conclusion regarding feedbacks, 48–51
 the most important ones in the atmosphere, **49**
 water vapour, 32, 43, 45, 48
Climate model, vii, 10, 12, 15, 17, 19, 21, 23, 25, 27, 32–34, 37, 41–43, 45, **46**, 47, 48, 79, 84, 105, 108, 137, 150, 157, 159, **161**
 Bern 2.5-dimensional model, ix, 8, **19**, 26, **28, 147**, 161, **162**
 comprehensive, 28
 coupled, 13, 16, 28, 42, 110, 140, **147, 148**, 146–150
 development (chronology), **14**
 hierarchy, viii, 2, 10, 25–33, 51, 79
 performance, **30–33**
 simplified, 19, 27
 the simplest (0 dimensions), 34
Climate modelling, vii, 7, 8, 15, 25, 29, 34, 45, 57, 67, 105, 118, 140, 146, 163, 164
 about deviations among the models, 29
 about the agreement with observations, 29
 components
  cloud cover, 32
  heat, 29

Index   173

precipitation, 29, 32
    water vapour, 29
  examples, 15–19
  historical development, 10–15
  purpose and limitations, 5–10
  role in climate science, **8**
Climate research, vii, 10, 13, 42, 58, 163
Climate science, vii, 2, 5, 8, 15, 53
  transition from descriptive to quantitative science, 7
Climate sensitivity, 8, 18, **51**, 41–51
  equilibrium, 41, 42, 48, 49, 51
  parameter, 42
  quantitative statements, 41
  water vapour, 45
Climate system, vii, 3–5, 15, 21, 22, **28**, 41, **44**, 45, 55, 57, 79, 151, 160–163
  components, **4**, 3–5, 150
Cloud cover, 32, 36, 43, 45–47
  change in radiation, **47**
  meridional distribution, **32**
Cloud feedback, 43, **46**, 45–48
CMIP, see Coupled modelling intercomparison project
Computational mode, see Numerical mode
Conservation equation, 98, 118
Continuity equation, 11, 56, 57, 101, 112, 118
Conveyor belt, oceanic, 85, 157
Coriolis
  acceleration, 99
  force, 12, 13, 87, 98–100, 110, 116, 125
  parameter, 99, 113
Coupled climate model, 13, 16, 28, 42, 110, 140, **147**, **148**, 146–150
Coupled climate-carbon cycle modelling intercomparison project (C^4MIP), 25
Coupled modelling intercomparison project (CMIP), 25
Coupling of physical model components, 137
Courant–Friedrichs–Lewy criterion (CFL criterion), 64–68, 70, 71, 73, 75, 95
Cryosphere, 4, 146
CTCS, see Numerical scheme, centered in time, centered in space

D/O events, see Dansgaard–Oeschger events
Dansgaard, Willy, 151
Dansgaard–Oeschger events, 151, **154**, 155, **155**
Dating, 6
Deep water circulation
  Atlantic, 160

Deep water formation, 6, 12, 121
  Atlantic, 152, 160
Density of sea water (as a function of temperature and salinity), 142
Derivative
  Eulerian, 98
  Lagrangian, 98
  material, 97–98
  partial, 98
  total, 97
Deterministic chaos, 13, 128
Diabatic heating, 126, 127, 129
Difference scheme, see Numerical scheme
Differential equation, 34, 37–41, 84, 106, 109
  ordinary (ODE), 37–41, 109, 128, 132
  partial (PDE), 1, 3, 13, 61, 62, 91, 94, 96, 105, 108, 109, 113, 126, 131
Diffusion, 27, 53–55
  eddy, 60
  molecular, 53–55, 60, 130
Diffusion constant/coefficient
  eddy, 60
  molecular, 54
  numerical, 73, 75, 77
Diffusivity, see Diffusion constant/coefficient
Drag, 126, 127
Driven flow
  thermally-driven, 125, 128
  wind-driven, 110–117, 120
Dronning Maud Land (Antarctica), 155

E-grid, 106, **107**
Earth as a rotational ellipsoid, 99
Earth Radiation Balance Experiment (ERBE), 35
Earth system, vii, 1
Earth system model of intermediate complexity (EMIC), viii, 10, 15, **19**, **26**, 27, 28, **28**, 85, 140, 146, **147**, 160
Easterlies, 111
EBM, see Energy balance model
Eddy, 58, 85, 87, 127, 145
  covariance, 60, 82
  diffusion, 60
  diffusion constant, 60
  diffusive flux density
    examples, **61**
  diffusivity, 84, 87, 88, **89**
  fluctuations, 58, **58**, 59, 60, 84
  flux, 88, **88**, 118, 127
  flux density, 60, 61
  heat flux, 127
  momentum flux, 127

role in mixing water masses, 89
stationary (SE), 82, **82**, 125
transient (TE), 82, **82**, 109, 125, 126
transport of angular momentum, 125
viscosity, xvii, 88, **89**
Ekman circulation, 87
El Niño-Southern Oscillation (ENSO), 19, 21
1997/1998, 20
2002/2003, 19–21
Electronic Numerical Integrator and Computer (ENIAC), 12
EMIC, *see* Earth system model of intermediate complexity
Emission scenario (SRES), 16, 27, 50, 162
Emissivity, 34, 36, 84
Energy
internal, 80, **81**, 82
kinetic, 80, **81**
latent, 80, **81**
potential, 80, **81**
total, 79, **81**
Energy balance model (EBM), 21, 25, 26, **26**, 27, 34, 36–38, 40, 42–44, **44**, 47, 84, 91
Budyko–Sellers, 84
meridional, 83–85
point model, 25, 34–37, 56
schematic depiction of two simple models, **35**
Energy flux, 5, 34, 53, 81, 82, 84, 107, 143
global, **7**
Energy transport, 79–89
heat, 79, **80**, 110, 157
atmosphere, 79–85
northern hemisphere, 79
ocean, 85–89
total, 83
ENIAC, *see* Electronic Numerical Integrator and Computer
Ensemble simulation, 15, 18, 25, 27
ENSO, *see* El Niño-Southern Oscillation
EPICA, *see* European Project for Ice Coring in Antarctica
Equation of motion, 98–101
Equilibrium climate sensitivity, 41, 42, 48, 49, 51
ERBE, *see* Earth Radiation Balance Experiment
Eulerian derivative, 98
European Project for Ice Coring in Antarctica (EPICA), 155
Evaporation, 5, 138, 142, 157, 160
leads to an increase of salinity, 142
temperature dependence, 138, 140

f-plane, *see also* β-plane, 99, **99**, 129
Faint young Sun, 156
FCT, *see* Numerical scheme, flux-corrected transport
Feedback, *see* Climate feedback
Feedback parameter, 42–44
Ferrel cell, 127, **127**, 128
Fick's first law, 54, 60, 145
Finite difference scheme, *see* Numerical scheme
Flux
correction, 13, 28, 110, 146, 148, **148**, 150, 160
energy, 5, 34, 53, 81, 82, 84, 107, 143
global, **7**
heat, 13, 82, 84, 85, 123, 126, 127, 131, 137, 138, 140, 145, 148, 152, 157
Atlantic, 138, **139**
correction, **149**
global, 138, **139**
meridional, 84
Newtonian, 140
restoring, 140, 141
internal energy, 82
mass, 121
matter, 53
momentum, 111, 126–128, 137, 144–145, 148
radiative, 37
restoring, **141**
salt, 143
restoring, 143
sensible heat, 138
water, 13, 137, 143, 145, 146, 148, 157
Atlantic, **144**
correction, **149**
global, **144**
Flux density
advective, 55
examples, **56**
eddy diffusive
examples, **61**
heat, 84
molecular diffusive, 53–55
examples, **55**
solar, 5
Flux quantity, 107, 108
Flux-corrected transport (numerical scheme), 77
Force
apparent (inertial), 98, 116
centrifugal, 98–100
Coriolis, 12, 13, 87, 98–100, 110, 116, 125
friction, 87, 98, 100, **100**

gravity, 98, 100
inertial (apparent), 98, 116
pressure gradient, 87, 98, 100, **100**, 116, 117, 120
real, 98, 100, 120
Forcing, 8, 42, 43
changes in land cover, 8
changes in solar radiation, 8
climate-relevant tracers, 8
cloud cover, 46, 47
doubling of atmospheric CO_2, 49
external, 5
radiative, 46, 49
volcanic eruptions, 8
Forward difference, *see* Numerical scheme, Euler forward difference
Fossil energy source, 19
Fourier expansion, 131
Free-fall acceleration, 100
Friction, 87, 98, 100, **100**, 111–113, 116, 120, 121, 126, 128
Frictionless ocean circulation, 120
FTCS, *see* Numerical scheme, Euler forward in time, centered in space
FTUS, *see* Numerical scheme, Euler forward in time, upstream in space
Fundamental laws (of physics and chemistry), 1

Gauss–Seidel method, 95, 96
General circulation, 123
atmospheric, 123–135
eddy fluxes of heat, **128**
mean wind field, **124**
meridional, **129**
oceanic, 97–122
Geo-engineering, 21, 23
Geophysical fluid dynamics, 1
Geostrophic equilibrium, **26**, 116, 122
Glacial period, 45, 85, 151, 160
cycles, 26
last, 151, 160
last maximum, 25
Global radiation, 5
long-wave, 5
short-wave, 5
Goal of the global community (1992), 22
Gradient operator, 54
Gravitational waves, 12
Gravity
acceleration, 80, 99
force, 98, 100
Greenhouse

effect, 5, 36, 37, 45, 47, 163
run-away, 28
gas, vii, 5, 15, **16**, 21, 22, 27, 33, 42, 43, 50, 160, 162, 163
anthropogenic, 3, 15
primary natural, 45
Greenland, 153, **155**, 163
abrupt climate changes, 151
ice core, 6, 7, 21, 27, 151, **152**, 153, 155
ice sheet, 21, 163
fate, 163
melting, 160
temperature, 153, 154, **154**
Greenland ice core project (GRIP), **152**, 155
Grid, 105–108
Arakawa (A-, E-, C-grid), 106, **107**
one-dimensional energy balance model, **108**
grid point, 38, 63, 71, 72, 95, 105, 106, 115
staggered, 106
GRIP, *see* Greenland ice core project
Gulf stream, 110, **149**
Gyre
barotropic, 86
subpolar, 110
subtropical, 110

Hadley
cell, 125, 127, **127**, 128
circulation, 125, **125**
Hadley, George, 125
Heat
capacity, 80, 82, 141, 142
equation, 68
flux, 13, 82, 84, 85, 123, 126, 127, 131, 137, 138, 140, 141, 145, 148, 152, 157
Atlantic, 138, **139**
correction, **149**
global, 138, **139**
flux density, 84
latent, 47, 80, 126, 127, 142
Helmholtz equation, 92
Hierarchy of coupled models, **26**
Holocene (present interglacial period), 151, 155
Humidity, 80, 143
Hydrological boundary condition, 142–143
Hydrological cycle, 158, 163
Hydrosphere, 3
Hydrostatic equation, 101
Hydrostatic equilibrium, 101, 102, 115, 118
Hydrostatic pressure, 102

Ice
-albedo feedback, 43–45, 50, 156
parameterisation, 156
core, 7, 151, 153, 155
about two pioneers in ice core science, 151
Antarctica, 6, 21, 151, 153, 155
Greenland, 21, 27, 151, 153, 155
sheet (North Atlantic), 160
Ideal gas law, 11
Implicit trapezoidal numerical scheme, 71, **71**, **73**
Inertial force, 98, 116
Infinite heat capacity model, 141
Initial condition, 12, 13, 15, 38, 40, 61, 62, 64, 66, 72, 91
in Richardson's first prediction, 12
Initial value problem, 91–96
Interglacial period, 45, 85, 163
Holocene, 151, 155
Intergovernmental Panel on Climate Change (IPCC), 16, 32, 160
Internal energy flux, 82
Intertropical Convergence Zone, 125
IPCC, *see* Intergovernmental Panel on Climate Change
ITCZ, *see* Intertropical Convergence Zone

Jacobi method, 95, 96
Jet stream, 67, 123, 125

Kelvin waves, 105
Kinematic viscosity, xvii, 130
Kuroshio current, 110, **149**
Kyoto-Protocol, 19

Lagrangian derivative, 98
Land surface, 4
Laplace
equation, 92
operator, 105
eigenfunctions on a sphere, 109
Lapse rate feedback, 47–48, 50
in the tropics and the mid-latitudes, **48**
Large-scale
atmospheric circulation, 123–135
eddy fluxes of heat, **128**
mean wind field, **124**
meridional, **129**
zonal and meridional, 123–128
flow, 13, 137

ocean flow, 118
oceanic circulation, 97–122
Latent heat, 47, 126, 127, 142
specific, 80, 143
Law of Newton, second, 98
Leap-frog numerical scheme, 63, 64, **64**, 68
explicit and implicit, 70, **71**
Legendre functions, 109
Lorenz, Edward, 11, 13, **13**, 84, 128
Lorenz–Saltzman model, 128–135
coordinates, solution domain, **130**
numerical solutions, **134**, **135**

Manabe, Syukuro, 13, **13**, **26**, 159
Marine isotope stage (MIS), 155
Mass balance equation, 57
Mass flux, 121
cross-isopycnal, 118
Material derivative, 97–98
Matter flux, 53
Mauna Loa (Hawai'i), CO_2 concentration, **22**
Mean
temporal, 58–60, 81, 86
zonal, 81, 86
Meridional energy flux, 82, **83**, 84
Meridional heat flux, 84, 85, **86–88**, 123, 152
Meridional heat flux density, 84
Meridional overturning circulation (MOC), 86, 87, 161, **161**, **162**
Atlantic, 155, 160, 162
Methane (CH_4), 5, **152**, 163
MIS, *see* Marine isotope stage
Mixed boundary condition, 145–146
MOC, *see* Meridional overturning circulation
Model
advection-diffusion, 27
atmosphere/ocean general circulation (AOGCM), 28, **31**, **50**
atmospheric general circulation (AGCM), **26**, 28, **29**, 32, **32**, 37, 85
circulation, 3, 12, 15
comparison, 33
Earth system model of intermediate complexity (EMIC), viii, 10, 15, **19**, **26**, 27, 28, **28**, 85, 140, 146, **147**, 160
general circulation, 13, 85
hydrological, 143
ocean general circulation (OGCM), **26**, 28, **29**, 87, **88**, 154
of Lorenz and Saltzman, 128–135

Index 177

of reduced complexity, see Earth system
  model of intermediate complexity
  (EMIC)
of Saltzman, 25
of Stommel, 110–117, 120, **121**, **122**
  numerical solution, **114**, **117**
  ocean basin, **111**
of Stommel and Arons, 121, **122**, **122**
pulse response, 26, 27
radiative-convective, 37
schematic illustration of 3-dimensional
  model grids, **29**
three-dimensional, 13, 26, 28, 79, 146, 151,
  160
Modelling intercomparison projects, 25
Molecular diffusion, 53–55, 60, 130
Molecular diffusive flux density, 53–55, 61
  examples, **55**
Momentum flux, 111, 126–128, 137, 144–145,
  148
Multiple equilibria, 146, 156–163
  coupled models, 159–163
  simple atmosphere model, 156–157
  simple ocean model, **159**, 157–159

Neural networks, 26, 27
Newtonian heat flux, 140
Newtons' second law, 98
Nitrogen, 21, 137
Nitrous oxide (N_2O), 5
Numerical diffusion, 76–77
  constant (diffusivity), 73, 75, 77
Numerical dispersion, 66, **66**
Numerical mode, 65, 66, **66**, 68, 69, **70**, 73, **73**
Numerical scheme, 67
  centered difference, 39, 40, 63, 67, 68, 106
  centered in time, centered in space (CTCS),
    63, 64, **64**, 68, 70, **71**
  Crank–Nicholson, 76
  Euler backward difference, 39, 69
  Euler difference, 38, 40, 41, **41**, 69
  Euler forward difference, 39, 63, 68, 69,
    72, 74, 94
  Euler forward in time, centered in space
    (FTCS), 64, 68, 72, 74
  Euler forward in time, upstream in space
    (FTUS), 69
  flux-corrected transport (FCT), 77
  implicit, **71**, 70–72
    trapezoidal, 71, **71**, **73**
  Lax, 72–74, **74**
  Lax–Wendroff, **74**, 74–75
  leap-frog, 63, 64, **64**, 68

  explicit and implicit, 70, **71**
  Runge–Kutta, 40, 41, **41**
  upstream, 69, 70
Numerical solution
  direct method, 92–94
  iterative method, 94–96
    Gauss–Seidel, 95, 96
    Jacobi, 95, 96
    relaxation, 94–95
    successive overrelaxation (SOR), 95–96
Numerical stability, 64–68

Ocean
  large-scale circulation, 97–122
Ocean carbon-cycle modelling inter-
  comparison project (OCMIP),
  25
Ocean general circulation model (OGCM), **26**,
  28, **29**, 87, **88**, 154
Ocean modelling intercomparison project
  (OMIP), 25
OCMIP, see Ocean carbon-cycle modelling
  intercomparison project
Oeschger Centre for Climate Change Research,
  University of Bern, vii
Oeschger, Hans, 151
OGCM, see Ocean general circulation model
OMIP, see Ocean modelling intercomparison
  project
Ordinary differential equation (ODE), 37–41,
  109, 128, 132

Paleo-data, viii, 7, 21
Paleo-reconstructions, 51
Paleo-thermometer, 6
Paleoclimate
  modelling intercomparison project (PMIP),
    25
  research, 28
  science, 7
Paleoclimatic archives, vii, 7, **9**, 15, 21, 151,
  155, 160, 163
parameterisation (used in climate models), 19,
  25, 27, 29, 32, 34, 84, 112, 137,
  138, 141, 144, 145, 150, 156
  Sellers' (temperature-dependence of the
    albedo), 44
  eddy flux density, 60
  energy transport, 79–89
  irradiance emitted from the Earth, 34
  temperature-dependence of the albedo, 44
  Sellers' parameterisation, 44

Partial differential equation (PDE), 1, 3, 13, 61, 62, 91, 94, 96, 105, 108, 109, 113, 126, 131
PCMDI, see Program for Climate Model Diagnosis and Intercomparison
Permafrost, 4, 6, 163
Pinatubo, volcanic eruption in 1991, 45, **46**
Planck feedback, 43, 48
Planetary waves, see Rossby waves
Planetary-gravity waves, 105
PMIP, see Paleoclimate modelling intercomparison project
Point model of the radiation balance, 34–37, 56
Poisson equation, 92
   direct numerical solution, 92–94
Potential vorticity, 117–122
Precipitation, 3, 20, 25, 29, 30, 32, 142, 145, 157
   decrease sea surface density, 160
   global annual mean, **33**
   meridional distribution, **31**
Pressure gradient force, 87, 98, 100, **100**, 116, 117, 120
Program for Climate Model Diagnosis and Intercomparison (PCMDI), 33
Pulse response model, 26, 27

Radiation
   back-, 5, 42
   balance, global, 5
   black body, 36
   grey body, 34, 36, 140
   long-wave, 34
   short-wave, 34
Radiation balance, as a function of latitude, 79, **80**
Radiative flux, 37
Radiative-convective model, 26
Real force, 98
Reconstruction of past climatic states, 5
Relative vorticity, 118
Relaxation method, 94–95
Remote sensing data, 20, 35, 48
Residual (methods of relaxation), 96
Restoring boundary condition, 143, 145, 146
Restoring flux, **141**
Restoring heat flux, 140, 141
Restoring salt flux, 143
Restoring temperature, 140, 141
Restoring time scale, 141
Richardson, Lewis Fry, 11, **11**, 12
Rossby, Carl-Gustav, 12

Rossby waves, 6, 12, **12**, 105

Salinity, 4, 85, 110, 142, 143, 145, 146, **147**, 157–159
   anomalies, **145**
Salt flux, 143
Saltzman, Barry, 128, 129
Saltzman model, 25
Scale-dependence, 142
Sea level rise, 160, 163
Sensible heat flux, 138
Sensitivity, see Climate sensitivity
Shallow water equations, 101–105
   a simplified version, 104, 105
   simple grid, **106**
   staggered grid, **106**
Shallow water model, **102**, 101–105, 118
   continuity equation, 104
   equations of motion, 102
   fundamental equations, 104
   one-dimensional, 105
      fundamental equations, 105
Shear stress, 100, **100**, 144
Snow-albedo feedback, 17
Snowball Earth hypothesis, 156
Soil moisture feedback, 17
Solar constant, xviii, 5, 15, 34, 84, 156
Solar flux density, 5
Solar radiation, xviii, 8, **16**, 37, 43, 125, 137, 138, 140
SOR, see Successive overrelaxation
Spatial scale, **6**
Specific heat
   capacity, 80, 82, 141
   latent, 80, 143
   of air, 34
Spectral methods, 108–110
   rhomboidal truncation, 109, **110**
   triangular truncation, 109, **110**
Spectral model, see Spectral methods
Spherical coordinates, 105
Spherical harmonics, 109
Spitsbergen, 110
SRES (Special Report on Emissions Scenarios), see Emission scenario
Stable isotopes
   in ice bubbles, 22
   of water, 6
      as a measure for temperature, 151
Staggered grid, 106
Stommel
   boundary layer, 113
   equation, 112

boundary conditions, 113
model, 110–117, 120, **121**, **122**
numerical solution, **114**, **117**
ocean basin, **111**
Stommel, Henry, **26**, 110–112, 121, 157
Stommel–Arons model, 121, **122**, **122**
Stream function, xvii, 112–113, **114**, 126, **127**, 129, **129**, **147**
Streamlines, 112
Successive overrelaxation (SOR), 95–96, 116
Summary for Policymakers (IPCC, Fourth Assessment Report of IPCC, 2007), 161
Symbols, list of, xvii

Taylor series, 37, 38, 76, 138, 139, 142
Teleconnection, 19, 20, 153
Temperature
 anomalies, **145**
 global and continental since 1900, **16**
 global equilibrium, 35, 36, **36**, 37, 42
 meridional change in the atomsphere over the last 20 years for emission scenario A1B, **50**
 meridional distribution, **31**, 123
 northern hemisphere, 8
 northern hemisphere over the last 1,200 years, **9**
 over Europe in summer 2070–2100, **18**
 since 1860, **9**
Termination I, 151, 155
THC, *see* Thermohaline circulation
Thermal boundary condition, 137–142
Thermal diffusivity, 131
Thermally-driven flow, 125, 128
Thermodynamic energy equation, 126
Thermohaline circulation (THC), 86–88, 151, **157**, 158, 159, **159**, 161
 "Atlantic heat pump", 151
Time
 averaging, 58, 59, 81
 slices, 17
 step, 38
Time scale, **6**
Tipping point, 163
Trade winds, 125
Transatlantic drift current, 110
Transfer coefficient, 138–140, 143, 145
Tunable parameter, 35, 51, 153

Tuning (model parameter tuning), 35, 36, 51, 153

Upwelling, 121, 122, **122**

Viscosity
 eddy, xvii, 88, **89**
 kinematic, xvii, 130
Volcanic eruption, 5, 8, 15, 16, 21
 Pinatubo (1991), 45, **46**
Volume quantity, 107
Vortex, 119, **119**
Vorticity, 12, 118
 absolute, 120
 potential, 117–122
 relative, 118
Vorticity equation
 Lorenz–Saltzman model, 130

Water (H_2O), 5, 42
 flux, 13, 137, 143, 145, 146, 148, 157
  Atlantic, **144**
  correction, **149**
  global, **144**
 vapour, 3, 4, 29, 32, 36, 43, 45, 47, 48, 50, 160
 feedback, 32, 43, 45, 48
Wave equation (classical), 62, 76, 105
Waves, 71
 gravitational, 12
 harmonic dispersion-free, 105
 Kelvin, 105
 planetary-gravity, 105
 Rossby, 6, 12, **12**, 105
Westerlies, 111, 125, 127
Western boundary current, 67, 87, 113, 121, **121**, 122, **122**
Wind stress, 113, 114, 116, 117, 127
Wind-driven circulation, 86, 110–117, 120
Wind-driven flow, 110–117, 120

Younger Dryas cooling, 155

Zero heat capacity model, 142
Zonal averaging, 81